An Introduction to Petroleum Technology, Economics, and Politics

Scrivener Publishing
3 Winter Street, Suite 3
Salem, MA 01970

Publishers at Scrivener
Martin Scrivener (martin@scrivenerpublishing.com)
Phillip Carmical (pcarmical@scrivenerpublishing.com)

An Introduction to Petroleum Technology, Economics, and Politics

James G. Speight

Co-published by John Wiley & Sons, Inc. Hoboken, New Jersey, and Scrivener Publishing LLC, Salem, Massachusetts.
Published simultaneously in Canada.

For general information on our other products and services or for technical support, please contact our Customer Care Department within the United States at (800) 762-2974, outside the United States at (317) 572-3993 or fax (317) 572-4002.

Wiley also publishes its books in a variety of electronic formats. Some content that appears in print may not be available in electronic formats. For more information about Wiley products, visit our web site at www.wiley.com.

For more information about Scrivener products please visit www.scrivenerpublishing.com.

Cover design by

Library of Congress Cataloging-in-Publication Data:

ISBN 978-1-118-01299-4

Printed in the United States of America

10 9 8 7 6 5 4 3 2 1

Contents

Preface

Crude oil is the major source of fuel used in the modern world. It is generally in a liquid state and is recovered by drilling and pumping, after which it is transported in tankers and pipelines.

The crude oil sector is the largest and most dominant economic sector of business in the United States. It is increasingly apparent that responsibility for the long-term consequences of economic and technological development decisions is extremely important and the logic of looking after the short term and letting the future take care of itself seems harder and harder to justify or sustain. Solving the supply problem requires innovative approaches, economic attractiveness, environmental appeal, and social responsibility. Such a course of action calls for background knowledge of the character of crude oil and the various factors that contribute to the price.

In fact, the market price on which contractual arrangements are settled is full of unknowns, concealing everything about trends in the costs of production. In the surplus, represented by the difference between the sale price obtained and the costs incurred, there is the shareholders' expected return on their investment (ROI), which determines whether, when, and on what terms petroleum-pricing influences the market.

The economics of crude oil pricing is one of the most complex and variable mechanisms in the commodities market. It is affected by a host of different factors, and it can be extremely difficult to determine which factors have the greatest impact on the actual spot price at any given point in time. In the past year, the crude oil markets have been extremely volatile; there have been claims that they have been subject to manipulation but the evidence is sorely lacking. On the other hand, if manipulation does not drive the oil markets, there is no evidence either that it does not take place. The whole issue is so debatable that the true drivers of the market may never be known with any degree of certainty.

One way for scientists, engineers, and people of lesser technical backgrounds to remove themselves from the dark shadow of economic guesswork is to understand, to some extent, the technological and political factors that are involved in crude oil economics.

Crude oil prices behave much as any other commodity, with wide price swings in times of shortage or oversupply. The crude oil price cycle may extend over several years, responding to changes in demand as well as supply. Indeed, the economics of oil must take into account that it is a depleting non-renewable resource and the cost of extraction of a non-renewable resource depends not only on the current rate of production but also the amount of cumulative production. The poignant question that always remains relates to the lifetime of current crude oil reserves and whether there are years or decades of reserves remaining.

Many pundits believe that the projections of running out of oil are based on geology, not price. Every existing oil reservoir has more than half of the original oil in place — many with more. These are resources that we know exist; we know where they are and what the oil looks like. Much of the crude oil that is left is trapped in tiny pores and cannot be recovered by simple pumping, and more advanced, expensive procedures are necessary to recover the crude oil.

Another aspect of crude oil economics is the cost of refining. Refining high-sulfur crude oil also requires greater expenditures for energy. In fact, energy accounts for approximately half of the refinery cost. Refinery location is yet another variable. The closer a refinery is to the crude oil source and the demand, the lower the transportation costs. Otherwise, the refinery must factor in the added cost of getting the products to market. The ultimate variable in crude oil economics is the price of crude oil, along with crude oil quality. High viscosity, high-sulfur crude oil can cost up to one-third less than low viscosity, low-sulfur crude oil. However, because high-sulfur crude oil requires more processing, refineries that buy primarily cheap crude oil incur more fixed expenses for equipment and labor.

After decades of stable — even cheap — crude oil during the first three-quarters of the 20th century, the geopolitical upsets of the 1970s led to rapid surges in crude oil prices. In the past five years, these surges have been magnified with crude oil process topping $147 per barrel in the summer of 2008, after which prices seemed to stabilize at approximately $80 per barrel. However, instability within the oil producing nations has, at the time of writing (March 2011), caused a surge in crude oil price to a figure in excess of $100 per barrel. There are opinions that such prices surges are merely bubbles that will burst and oil prices will return to lower levels.

However, many economists are unable to explain the economics of crude oil pricing without recourse to higher mathematics. The result is the development of complex equations that are not only difficult to understand but also bear little relationship to reality. In fact, when oil prices flip-flop, explanations are invoked and justified, using the remarkable facet of 20/20 hindsight with very little foresight or even knowledge of the workings of the industry.

This book will introduce the reader to the factors that influence the price of crude oil insofar as crude oil economics involves a combination of several factors, not the least of which are:

1. Crude oil availability from the reservoir
2. Crude oil quality
3. Crude oil extraction
4. Crude oil quality
5. Geopolitics.

The book also includes the reader to oil classification, recovery, and properties, which are not usually included in works related to politics and economics but are an essential part of for understanding oil pricing and politics.

Dr. James G. Speight
Laramie, Wyoming
March 2011

1

History and Terminology of Crude Oil

Geology and time have created reservoirs of crude oil (petroleum) in various parts of the world. Until the mid-1800s, this vast untapped wealth lay mostly hidden below the surface of the earth. Some oil naturally seeped to the earth's surface and formed shallow pools that were used as a source of medicinal liquids, illuminating oil, and, after evaporation of the volatile components, as a caulking for boats and a building mastic (Speight, 2007). For centuries, demand was limited but better refining techniques and surging demand for kerosene and lubricants in the late 19th century changed this.

Crude oil is the major source of fuel used by people today. Because crude oil is liquid, it is easy to recover by drilling and pumping, rather than excavation, and it is easy to transport in tankers and pipelines. In fact, the rapid rise in crude prices in the past years has strengthened calls for renewed initiatives on energy security for petroleum importing countries. While there has been a convergence of factors contributing to the current high oil prices, oil supply and demand fundamentals, the role of speculative forces, and structural bottlenecks in the downstream sector have emerged as the main areas of concern.

The demand for gasoline and middle distillates (including aviation fuels) has risen significantly while refining capacity has only shown a modest increase, if any. This growth in demand over and above the increase in refining capacity has significantly raised refinery utilization rates and tightened the downstream market, raising serious concerns over a potential supply gap in the downstream oil market. This issue is particularly prevalent in the United States, where low surplus refining capacity and stringent oil product specifications have resulted in reduced flexibility in the refining sector to adjust to changes in seasonal demand patterns.

The economics of oil must take into account that it is a depleting non-renewable resource and the cost of extraction of a non-renewable resource depends not only on the current rate of production but also on the amount of cumulative production. Crude oil prices behave much as any other commodity with wide price swings in times of shortage or oversupply. The crude oil price cycle may extend over several years responding to changes in demand as well as supply. Many pundits believe that the projections of running out of oil are based on geology, not price. Every existing oil reservoir has more than half of the original oil in place, many with more. These are resources that we know exist; we know where they are and what the oil looks like. Much of the crude oil that is left is trapped in tiny pores and cannot be recovered by simple pumping, and more advanced and expensive procedures are necessary to recover the crude oil.

Another aspect of crude oil economics is the cost of refining. Refining high-sulfur crude oil also requires greater expenditures for energy. In fact, energy accounts for approximately half of the refinery cost. Refinery location is yet another variable. The closer a refinery is to the crude oil source and the demand, the lower the transportation costs . Otherwise, the refinery must factor in the added cost of getting the products to market. Obviously, the ultimate variable in crude oil economics is the price of crude oil. Crude oil quality is another key variable. High viscosity, high-sulfur crude oil can cost up to one-third less than low viscosity, low-sulfur crude oil. However, because high-sulfur crude oil requires more processing, refineries that buy primarily cheap crude oil incurs more fixed expenses for equipment and labor.

While there is a growing need to address these issues, there exist barriers and constraints to the older oil person and the neophyte alike, as well as the economist. Often the terminology employed by the industry is so confusing that the ensuing issues and the issues involved in oil pricing and oil product pricing are a mystery. In addition, many economists are unable to explain the economics of oil and oil product pricing without recourse to higher mathematics. The result is the development of complex equations that are often difficult to understand, and, for the technical person in industry, appear to bear little relationship to what he understands in terms of oil properties.

Thus, it is appropriate to commence this book with a description of the historical uses of crude oil and crude oil terminology, leading the reader to a better understanding of the terminology of crude oil and the means by which it is described.

1.1 Historical Perspectives

Petroleum is the most important raw material used in modern society insofar as it provides not only raw materials for fuel manufacture and energy, but also starting materials for plastics and other products.

The word *petroleum*, derived from the Latin *petra* and *oleum*, literally means *rock oil* and refers to hydrocarbons that occur widely in the sedimentary rocks in the form of gases, liquids, semisolids, or solids. From a chemical standpoint, petroleum is an extremely complex mixture of hydrocarbon compounds, usually with minor amounts of nitrogen-, oxygen-, and sulfur-containing compounds, as well as varying amounts of metal-containing compounds (Speight, 2007).

The fuels that are derived from petroleum supply more than half of the world's total supply of energy. Gasoline, kerosene, and diesel oil provide fuel for automobiles, tractors, trucks, aircraft, and ships. Fuel oil and natural gas are used to heat homes and commercial buildings, as well as to generate electricity. Petroleum products are the basic materials used for the manufacture of synthetic fibers for clothing and in plastics, paints, fertilizers, insecticides, soaps, and synthetic rubber. The uses of petroleum as a source of raw material in manufacturing are central to the functioning of modern industry.

Petroleum is a carbon-based resource, so the geochemical carbon cycle is also of interest to fossil fuel usage in terms of petroleum formation, use, and the buildup of atmospheric carbon dioxide. Thus, the more efficient use of petroleum is of paramount importance. Petroleum technology, in one form or another, will be with us until suitable alternative forms of energy are readily available (Boyle, 1996; Ramage, 1997). Therefore, a thorough understanding of the benefits and limitations of petroleum recovery and processing is necessary, and hopefully can be introduced within the pages of this book.

The history of any subject is the means by which the subject is studied in the hopes that much can be learned from the events of the past. In the current context, the occurrence and use of petroleum, petroleum derivatives (naphtha), heavy oil, and bitumen is not new. The use of petroleum and its derivatives was practiced in pre-Christian times and is known largely through historical use in many of the older civilizations (Henry, 1873; Abraham, 1945; Forbes, 1958a, 1958b; James and Thorpe, 1994; Krishnan and Rajagopal, 2003). Thus, the use of petroleum and the development of related technology is not such a modern subject as we are inclined to believe. However, the petroleum industry is essentially a 20th century industry but to understand the evolution of the industry, it is essential to have a brief understanding of the first uses of petroleum.

The Tigris-Euphrates valley, in what is now Iraq, was inhabited as early as 4000 BC by the people known as the Sumerians, who established one of the first great cultures of the civilized world. The Sumerians devised the cuneiform script, built the temple towers known as ziggurats, an impressive law, literature, and mythology. As the culture developed, bitumen or asphalt was frequently used in construction and in ornamental works.

Although it is possible to differentiate between the words *bitumen* and *asphalt* in modern use, the occurrence of these words in older texts offers no such possibility. It is significant that the early use of bitumen was in the nature of cement for securing or joining together various objects, and it therefore seems likely that the name itself was expressive of this application.

The word *asphalt* is derived from the Akkadian term *asphaltu* or *sphallo*, meaning *to split*. It was later adopted by the Homeric Greeks in the form of the adjective ασφαλής ες signifying *firm, stable, secure*, and the corresponding verb ασφαλίζω ίσω meaning *to make firm or stable, to secure*. Just like bitumen, the first use of asphalt by the ancients was in the nature of cement for securing or joining together various objects, such as the bricks used for building, so it seems likely that the name itself was also expressive of this application. From the Greek, the word passed into Latin (*asphaltum, aspaltum*), and thence into French (*asphalte*) and English (*aspaltoun*).

The origin of the word *bitumen* is more difficult to trace and subject to considerable speculation. The word was proposed to have originated in the Sanskrit, where we find the words *jatu*, meaning *pitch*, and *jatukrit*, meaning *pitch creating*. From the Sanskrit, the word *jatu* was incorporated into the Latin language as *gwitu* and is believed to have eventually become *gwitumen* (pertaining to pitch). Another word, *pixtumen* (exuding or bubbling pitch) is also reputed to have been in the Latin language, although the construction of this Latin word form from which the word *bitumen* was reputedly derived, is certainly suspect. There is the suggestion that subsequent derivation of the word led to a shortened version, which eventually became the modern version, called *bitumen*, thence passing via French into English. From the same root is derived the Anglo Saxon word *cwidu* (mastic, adhesive), the German work *kitt* (cement or mastic) and the equivalent word *kvada*, which is found in the old Norse language as being descriptive of the material used to waterproof the long ships and other sea-going vessels. It is just as likely that the word is derived from the Celtic *bethe* or *beithe* or *bedw* that was the birch tree that was used as a source of resin. The word appears in Middle English as *bithumen*. In summary, a variety of terms exist in ancient language that from their described use in texts can be proposed as meaning bitumen or asphalt (Abraham, 1945).

Using these ancient words as a guide, it is possible to trace the use of petroleum and its derivatives as described in ancient texts. And, preparing derivatives of petroleum was well within the area of expertise of the early scientists since alchemy (early chemistry) was known to consist of four sub-routines: dissolving, melting, combining, and distilling (Cobb and Goldwhite, 1995).

Early references to petroleum and its derivatives occur in the Bible, although by the time the various books of the Bible were written, the use of petroleum and bitumen was established and it is apparent that bitumen and petroleum derivatives were items of commerce. The exact prices paid are unknown and may even have been given as a tribute to the local king. Nevertheless, in spite of the missing monetary values, these writings do offer documented examples of the use and trade of petroleum and related materials.

For example, in the Epic of Gilgamesh written more than 2,500 years ago, a great Flood causes the hero to build a boat that is caulked with bitumen and pitch (see for example, Kovacs, 1990). And, in a related story of Mesopotamia, just prior to the flood,

Noah is commanded to build an ark that also includes instructions for caulking the vessel with pitch (Genesis 6:14):

> Make thee an ark of gopher wood; rooms shalt thou make in the ark, and shalt pitch it within and without with pitch.

The occurrence of *slime* (bitumen) *pits* in the Valley of Siddim (Genesis, 14:10), a valley at the southern end of the Dead Sea, is reported. There is also reference to the use of tar as a mortar when the Tower of Babel was under construction (Genesis 11:3):

> And they said one to another, Go to, let us make brick, and burn them thoroughly. And they had brick for stone, and slime had they for mortar.

In the Septuagint, or Greek version of the Bible, this work is translated as *asphaltos*, and in the Vulgate or Latin version, as *bitumen*. In the Bishop's Bible of 1568 and in subsequent translations into English, the word is given as *slime*. In the Douay translation of 1600, it is *bitume*, while in Luther's German version, it appears as *thon*, the German word for clay.

Another example of the use of pitch (and slime) is given in the story of Moses (Exodus 2:3):

> And when she could no longer hide him, she took for him an ark of bulrushes, and daubed it with slime and with pitch, and put the child therein; and she laid it in the flags by the river's brink.

Perhaps the slime was a lower melting bitumen (bitumen mixed with solvent) whereas the pitch was a higher melting material; the one (*slime*) acting as a flux for the other. The lack of precise use of the words for bitumen and asphalt as well as for tar and pitch even now makes it unlikely that the true nature of the biblical tar, pitch, and slime will ever be known, but one can imagine their nature. In fact, even modern Latin dictionaries give the word bitumen as the Latin word for asphalt.

It is most probable that, in both these cases, the pitch and the slime were obtained from the seepage of oil to the surface, which was a fairly common occurrence in the area. And during biblical times, bitumen was exported from Canaan to various parts of the countries that surround the Mediterranean (Armstrong *et al.*, 1997).

In terms of liquid products, there is an interesting reference (Deuteronomy, 32:13) to bringing oil out of flinty rock. The exact nature of the oil is not described nor is the nature of the rock. The use of oil for lamps is also referenced (Matthew, 23:3), but whether it was mineral oil (a petroleum derivative such as naphtha) or whether it was vegetable oil is not known.

Excavations conducted at Mohenjo-Daro, Harappa, and Nal in the Indus Valley indicated that an advanced form of civilization existed there. An asphalt mastic composed of a mixture of asphalt, clay, gypsum, and organic matter was found between two brick walls in a layer about 25 mm thick — probably a waterproofing material. Also unearthed was a bathing pool that contained a layer of mastic on the outside of its walls and beneath its floor.

In the Bronze Age, dwellings were constructed on piles in lakes close to the shore to better protect the inhabitants from the ravages of wild animals and attacks from marauders. Excavations have shown that the wooden piles were preserved from decay by a coating of asphalt, and posts preserved in this manner have been found in Switzerland. There are also references to deposits of bitumen at Hit (the ancient town of Tuttul on the Euphrates River in Mesopotamia) and the bitumen from these deposits was transported to Babylon for use in construction (Herodotus, *The Histories*, Book I). There is also reference to a Carthaginian story in which birds' feathers smeared with pitch are used to recover gold dust from the waters of a lake (Herodotus, *The Histories*, Book IV).

One of the earliest recorded uses of asphalt was by the pre-Babylonian inhabitants of the Euphrates Valley in southeastern Mesopotamia, present-day Iraq, formerly called Sumer and Akkad and, later, Babylonia (Thompson, 1936; Moorey, 1994). In this region there are various asphalt deposits and the uses of the material have become evident. For example, King Sargon of Akkad (Agade c. 2550 BC) was set adrift by his mother in a basket of bulrushes on the waters of the Euphrates, where he was discovered by Akki the husbandman (irrigator), who brought him up to serve as gardener in the palace of Kish. Sargon eventually ascended to the throne.

On the other hand, the bust of Manishtusu, King of Kish, an early Sumerian ruler (about 2270 BC), was found in the course of excavations at Susa in Persia, and the eyes, composed of white limestone, are held in their sockets with the aid of bitumen.

Fragments of a ring composed of asphalt have been unearthed above the flood layer of the Euphrates at the site of the prehistoric city of Ur in southern Babylonia, ascribed to the Sumerians of about 3500 BC.

An ornament excavated from the grave of a Sumerian king at Ur consists of a statue of a ram with the head and legs carved out of wood over which gold foil was cemented by means of asphalt. The back and flanks of the ram are coated with asphalt in which hair was embedded. Another art of decoration consisted of beating thin strips of gold or copper, which were then fastened to a core of asphalt mastic. An alternative method was to fill a cast metal object with a core of asphalt mastic, and such specimens have been unearthed at Lagash and Nineveh. Excavations at Tell-Asmar, 50 miles northeast of Baghdad, revealed the use of asphalt by the Sumerians for building purposes.

Mortar composed of asphalt has also been found in excavations at Ur, Uruk, and Lagash, and excavations at Khafaje have uncovered floors composed of a layer of asphalt that has been identified as asphalt, mineral filler (loam, limestone, and marl), and vegetable fibers (straw). Excavations at the city of Kish (Persia) in the palace of King Ur-Nina showed that the foundations consist of bricks cemented together with an asphalt mortar. Similarly, in the ancient city of Nippur (about 60 miles south of Baghdad), excavations show Sumerian structures composed of natural stones joined together with asphalt mortar. Excavation has uncovered an ancient Sumerian temple in which the floors are composed of burnt bricks embedded in asphalt mastic that still shows impressions of reeds with which it must have been originally mixed.

The Epic of Gilgamesh, written before 2500 BC and transcribed on to clay tablets during the time of Ashurbanipal, king of Assyria (668 to 626 BC), make reference to the use of asphalt for building purposes. In the eleventh tablet, Ut-Napishtim relates the well-known story of the Babylonian flood, stating that "he "smeared the" inside of a boat with six sar of kupru and the outside with three sar…"

Kupru may have meant that the pitch or bitumen was mixed with other materials (perhaps even a solvent such as distillate from petroleum) to give it the appearance of slime as mentioned in the Bible. In terms of measurement, *sar* is a word of mixed origin and appears to mean an interwoven or wickerwork basket. An approximate translation is:

> The inside of the boat was smeared (coated, caulked) with six baskets full of pitch and the outside of the boat was smeared (coated, caulked) with three baskets full of pitch.

There are also indications from these texts that that asphalt mastic was sold by volume, or by the gur. On the other hand, bitumen was sold by weight — by the mina or shekel — for which there is no recorded price, although the word *shekel* is often used to mean a particular type of coinage as well as the weight of a commodity. In fact, the use of petroleum in seepages and bitumen from surface tar sand deposits seems to have been widely used and it is very likely that there was a ready trade for the raw material and the products made from petroleum and bitumen were actually looked upon as commodities (Barton, 1926; Sperber, 1976; Steinsaltz, 1977; Stern *et al.*, 2007). This is a situation that did not occur again until the 1980s (Chapter 5).

For example, use of asphalt by the Babylonians (1500 to 538 BC) is also documented. In fact, the Babylonians were well versed in the art of building, and each monarch commemorated his reign and perpetuated his name by the construction of a building or other monuments. The use of bitumen mastic as a sealant for water pipes, water cisterns, and in outflow pipes leading from flush toilets cities such as Babylon, Nineveh, Calah, and Ur has been observed and the bitumen lines are still evident (Speight, 1978).

Bitumen was used as mortar from very early times, and sand, gravel, or clay was employed in preparing these mastics. Asphalt-coated tree trunks were often used to reinforce wall corners and joints, for instance, in the temple tower of Ninmach in Babylon. In vaults or arches, a mastic-loam composite was used as mortar for the bricks, and the keystone was usually dipped in asphalt before being set in place. The use of bituminous mortar was introduced into the city of Babylon by King Hammurabi, but the use of bituminous mortar was abandoned toward the end of Nebuchadnezzar's reign in favor of lime mortar to which varying amounts of asphalt were added. The Assyrians recommended the use of asphalt for medicinal purposes, as well as for building purposes, and there is some merit in the fact that the Assyrian moral code recommended that asphalt, in the molten state, be poured onto the heads of delinquents. Pliny, the Roman author, also notes that bitumen could be used to stop bleeding, heal wounds, drive away snakes, treat cataracts as well as a wide variety of other diseases, and straighten out eyelashes that inconvenience the eyes. One can appreciate the use of bitumen to stop bleeding, but its use to cure other ailments is questionable and one has to consider what other agents were being used concurrently with bitumen.

The Egyptians were the first to adopt the practice of embalming their dead rulers and wrapping the bodies in cloth.

Before 1000 BC, asphalt was rarely used in mummification, except to coat the cloth wrappings and thereby protect the body from the elements. After the viscera had been removed, the cavities were filled with a mixture of resins and spices, the corpse immersed in a bath of potash or soda, dried, and finally wrapped. From 500 to about 40 BC, asphalt was generally used both to fill the corpse cavities and to coat the cloth wrappings. The word *mûmûia* first made its appearance in Arabian and Byzantine literature about 1000 AD, signifying bitumen. It is believed that it was the spread of the Islamic Empire that brought Arabic science and the use of bitumen to Western Europe.

In Persian, the term bitumen is believed to have acquired the meaning equivalent to paraffin wax that may be symptomatic of the nature of some of the crude oils in the area. Alternatively, it is also possible that the destructive distillation of bitumen to produce pitch produced paraffins that crystallized from the mixture over time. In Syriac, the term alluded to substances used for mummification. In Egypt, resins were used extensively for purposes of embalming up to the Ptolemaic period, when asphalts gradually came into use.

Arabian physician Al Magor used Mûmûia in prescriptions for the treatment of contusions and wounds as early as the 12th century. Its production soon became a special industry in the Alexandria. The scientist Al-Kazwînî alluded to the healing properties of mûmûia, and Ibn Al-Baitâr gives an account of its source and composition. In his treatise *Amoenitates Exoticae*, Engelbert Kämpfer (1651–1716) gives a detailed account of the gathering of mûmûia, the different grades and types, and its curative properties in medicine. As the supply of mummies was limited, other expedients came into vogue. The corpses of slaves or criminals were filled with asphalt, swathed, and artificially aged in the sun. This practice continued until the French physician, Guy de la Fontaine, exposed the deception in 1564 AD.

Many other references to bitumen occur throughout the Greek Empire and the Roman Empire, and from then to the Middle Ages, early scientists frequently alluded to the use of bitumen. In later times, both Christopher Columbus and Sir Walter Raleigh (depending upon the country of origin of the biographer) have been credited with the discovery of the asphalt deposit on the island of Trinidad and apparently used the material to caulk their ships.

The use of petroleum has also been documented in China; as early as 600 BC, petroleum was encountered when drilling for salt and mention of petroleum as an impurity in the salt is also noted in documents of the third century AD (Owen, 1975). It is presumed that the petroleum that contaminated the salt may be similar to that found in Pennsylvania and was, therefore, a more conventional type rather than the heavier type.

There was also an interest in the thermal product of petroleum (nafta; naphtha) when it was discovered that this material could be used as an illuminant and as a supplement to asphalt incendiaries in warfare. For example, there are records of the use of mixtures of pitch and/or naphtha with sulfur as a weapon of war during the Battle of Palatea, Greece, in the year 429 BC (Forbes, 1959). There are references to the use of a liquid material, naft (presumably the volatile fraction of petroleum, which we now call naphtha and is used as a solvent or as a precursor to gasoline as an incendiary material during various battles of the pre-Christian era (James and Thorpe, 1994). This is the so-called Greek fire, a precursor and chemical cousin to napalm. Greek fire is also recorded as being used in the period 674 to 678 when the city of Constantinople was saved by the use of the fire against an Arab fleet (Davies, 1996). In 717 to 718 AD, Greek fire was again used to save the city of Constantinople from attack by another Arab fleet, again with deadly effect (Dahmus, 1995). After this time, the Byzantine navy of three hundred triremes frequently used Greek fire against all comers (Davies, 1996). This probably represents the first documented use of the volatile derivatives of petroleum that led to a continued interest in petroleum.

Greek fire was a viscous liquid that ignited on contact with water and was sprayed from a pump-like device on to the enemy. One can imagine the early users of the fire attempting to ignite the liquid before hurling it towards the enemy.

However, the hazards that can be imagined from such tactics could become very real, and perhaps often fatal, to the users of the Greek fire if any spillage occurred before ejecting the fire towards the enemy. The later technology for the use of Greek fire most likely incorporated a heat-generating chemical such as quicklime (CaO) (Cobb and Goldwhite, 1995) that was suspended in the liquid and which, when coming into contact with water to produce $Ca(OH)_2$, released heat that was sufficient to cause the liquid to ignite. One can assume that the users of the fire were extremely cautious during periods of rain, or, if at sea, during periods of turbulent weather.

As an aside, the use of powdered lime in warfare is also documented. The English used it against the French on August 24, 1217 with disastrous effects for the French. As was usual for that time, there was a difference of opinion between the English and the French that resulted in their respective ships meeting at the east end of the English Channel. Before any other form of engagement could occur, the lime was thrown from the English ships and carried by the wind to the French ships, where it made contact with the eyes of the French sailors. The burning sensation in the eyes was too much for the French sailors and the English prevailed with the capture of much booty (Powicke, 1962).

The combustion properties of bitumen (and its fractions) were known in Biblical times. There is the reference to these properties (Isaiah, 34:9) when it is stated:

> *And the stream thereof shall be turned into pitch, and the dust thereof into brimstone, and the land thereof shall become burning pitch.* It shall not be quenched night nor day; the smoke thereof shall go up forever: from generation to generation it shall lie waste; none shall pass through it forever and forever.

One might surmise that the effects of the burning bitumen and sulfur (brimstone) were long-lasting and quite devastating.

Approximately 2,000 years ago, Arabian scientists developed methods for the distillation of petroleum, which were introduced into Europe by way of Spain. This represents another documented use of the volatile derivatives of petroleum that led to a continued interest in petroleum and its derivatives as medicinal and warfare materials, in addition to the usual construction materials.

From 1271 to 1273, Marco Polo also reported the Baku region of northern Persia as having an established commercial petroleum industry. It is believed that the prime interest was in the kerosene fraction that was then known for its use as an illuminant. By inference, it can be concluded that the distillation, and perhaps the thermal decomposition, of petroleum were established technologies. If not, Polo's diaries may well have contained a description of the stills or the reactors.

In addition, bitumen was investigated in Europe during the Middle Ages (Bauer, 1546, 1556), and the separation and properties of bituminous products were thoroughly described. Other investigations continued, leading to a good understanding of the sources and use of this material even before the birth of the modern petroleum industry (Forbes, 1958a, 1958b).

There are also records of the use of petroleum spirit, probably a higher boiling fraction of or than naphtha that closely resembled the modern-day liquid paraffin, for medicinal purposes. In fact, liquid paraffin has continued to be prescribed up to modern times. The naphtha of that time was obtained from shallow wells or by the destructive distillation of asphalt.

Parenthetically, the destructive distillation operation may be likened to modern coking operations in which the overall objective is to convert the feedstock into distillates for use as fuels. This particular interest in petroleum and its derivatives continued with an increasing interest in nafta (naphtha) because of its aforementioned used as an illuminant and as a supplement to asphaltic incendiaries for use in warfare.

Finally, not wishing to omit the use of bitumen (and, perhaps, its recognition as a tradable commodity), in the Americas, Pre-Hispanic Mesoamerican peoples collected, processed, and used bitumen as a decoration, a sealant and an adhesive (Wendt and Lu, 2006). Among the earliest to do so were the Olmec people (flourished ca. 1200 to 500 BC) of the Southern Coastal Lowlands of Mexico. Furthermore, geochemical analyses of bitumen from the Olmec archeological sites indicate a multiple procurement network that reflected wide trade and interactions between the various groups. It is also believed that bitumen processing was an organized and specialized activity involving multiple production stages (Wendt and Cyphers, 2008).

To continue such references is beyond the scope of this book, although they do give a flavor of the developing interest in petroleum. However, it is sufficient to note that there are many other references to the occurrence and use of bitumen or petroleum derivatives up to the beginning of the modern petroleum industry (Cook and Despard, 1927; Mallowan and Rose, 1935; Nellensteyn and Brand, 1936; Mallowan, 1954; Forbes, 1958a, 1958b, 1959, 1964; Marschner *et al.*, 1978).

In summary, the use of petroleum and related materials has been observed for almost 6,000 years. During this time, the use of petroleum has progressed from the relatively simple use of asphalt from

Mesopotamian seepage sites to the present-day refining operations that yield a wide variety of products and petrochemicals (Speight, 2007).

1.2 Modern Perspectives

The modern petroleum industry began in the later years of the 1850s with its discovery in 1857 and its subsequent commercialization in Pennsylvania in 1859 (Bell, 1945). The modern refining era can be said to have commenced in 1862 with the first appearance of a commercial unit for the distillation of petroleum (Speight, 2007) (Chapter 5).

Benjamin Silliman Sr. had a degree in law but he was also as qualified for geology as he was to be Yale professor of chemistry. The geology venture prospered, and by 1820, Silliman was in great demand for field trips on which he took his son, Benjamin Silliman Jr. When Silliman Sr. retired in 1853, Silliman Jr. took up where father had left off, as professor of general and applied chemistry at Yale (this time, with a degree in the subject). After writing a number of chemistry books and being elected to the National Academy of Sciences, Silliman Jr. took up lucrative consulting posts with the Boston City Water Company and various mining enterprises.

In 1855, one of his clients asked him to research and report on some mineral samples from the new Pennsylvania Rock Oil Company. After several months of work, Silliman Jr. announced that about 50% of the black tar-like substance could be distilled into first-rate burning oils (which would eventually be called kerosene and paraffin) and that an additional 40% of what was left could be distilled for other purposes, such as lubrication and gaslight.

With the acquisition of the original lighting oil (whale oil) seeing an increased demand and becoming more dangerous to acquire and on the basis of this report, a company was launched to finance the drilling of the Drake Well at Oil Creek, Pennsylvania. In 1857, it became the first well to produce petroleum. It would be another fifty years before Silliman Jr.'s reference to the other fractions available from the oil through extra distillation would provide gasoline for the combustion engine of the first automobile. Silliman Jr.'s report changed the world because it made possible an entirely new form of transportation and helped turn the United States into an industrial superpower. But, back to the future.

After completion of the first well by Edwin Drake, the surrounding areas were immediately leased and extensive drilling took place. Crude oil output in the United States increased from approximately 2000 barrels (1 barrel, bbl = 42 US gallons = 35 Imperial gallons = 5.61 $foot^3$ = 158.8 liters) in 1859 to nearly 3,000,000 barrels in 1863 and approximately 10,000,000 barrels in 1874. In 1861 the first cargo of oil, contained in wooden barrels, was sent across the Atlantic to London, and by the 1870s, refineries, tank cars, and pipelines had become characteristic features of the industry, mostly through the leadership of Standard Oil that was founded by John D. Rockefeller (Johnson, 1997). Throughout the remainder of the 19th century, the United States and Russia were the two areas in which the most striking developments took place.

At the outbreak of World War I in 1914, the two major producers of petroleum were the United States and Russia, but supplies of oil were also being obtained from Indonesia, Rumania, and Mexico. During the 1920s and 1930s, attention was also focused on other areas for oil production, such as the United States, the Middle East, and Indonesia. At this time, European and African countries were not considered major oil-producing areas. In the post-1945 era, Middle Eastern countries continued to rise in importance because of new discoveries of vast reserves. The United States, although continuing to be the biggest producer, was also the major consumer and not a key exporter of oil.

In the United States, approximately 20% of domestic production currently comes from marginal wells, which are the most vulnerable to low prices. Since 1998, domestic production has dropped to approximately 5.5 million barrels/day, with approximately twice that amount being imported.

At this time, oil companies are beginning to roam much farther afield in the search for oil, and, as a result, there have been significant discoveries in several countries.

1.3 Oil Companies

The term oil company is taken to mean a company that deals with the exploration, recovery, and refining of oil. Furthermore, in the United States, the term "big oil company" is taken to mean the major private international oil companies, largely based in Europe or North America (Pirog, 2007).

However, while some of those companies are indeed among the largest in the world, by many important measures, a majority of the largest oil companies are state-owned, national oil companies. By conventional definitions, national oil companies hold the majority of petroleum reserves and produce the majority of the world's supply of crude oil. Because national oil companies generally hold exclusive rights to the exploration and development of petroleum resources within the home country, they also can decide on the degree to which they require participation by private companies in those activities.

Rankings of oil companies can be based on current production (Table 1.1) to generate current earnings, and several standards need to be applied to assess the evolving nature of the companies in the industry to ensure the future viability of the enterprise on reserve positions (Table 1.2). Investment, in the form of exploration and development expenditures, serves as an indicator of the potential reserve and production positions of an oil company.

Privately held companies have the goal of maximizing shareholder value. The management of the company may accomplish

Table 1.1 Comparative ranking of the top ten oil companies based on current liquids production.

Rank 2006	Company	Production	Rank 2000	Company	Production
1	Saudi Aramco	11,035	1	Saudi Aramco	8,044
2	NIOC	4,049	2	NIOC	3,620
3	Pemex	3,710	3	Pemex	3,343
4	PDV	2,650	4	PDV	2,950
5	KPC	2,643	5	INOC	2,528
6	BP	2,562	6	ExxonMobil	2,444
7	ExxonMobil	2,523	7	Shell	2,268
8	PetroChina	2,270	8	PetroChina	2,124
9	Shell	2,093	9	BP	2,061
10	Sonotrach	1,934	10	KPC	2,025

Table 1.2 Comparative ranking of the top ten oil companies based on current reserve holdings.

Rank 2006	Company	Reserves	Rank 2000	Company	Reserves
1	Saudi Aramco	264,200	1	Saudi Aramco	259,200
2	NIOC	137,500	2	INOC	112,500
3	NIOC	115,000	3	KPC	96,500
4	KPC	101,500	4	NIOC	87,993
5	PDV	79,700	5	PDV	76,852
6	Adnoc	56,920	6	Adnoc	50,710
7	Libya NOC	33,235	7	Pemex	28,400
8	NNPC	21,540	8	Libya NOC	23,600
9	Lukoil	16,114	9	NNPC	13,500
10	QP	15,200	10	Lukoil	11,432

that goal through organizing production so that a profit is made in the current timeframe as well as in the future. The management may also make investment decisions to take advantage of opportunities to raise the company's rate of return. They also have the motivation to achieve productive efficiency to hold down costs and enhance the profitability of any given revenue level. This activity is thought to benefit consumers by assuring that physical shortages are avoided and that the good is available at the lowest price consistent with demand and supply factors.

1.4 Definitions and Terminology

Throughout the previous sections, definitions and terminology have been represented that give some insight into the terminology of the industry. However, there are many aspects of petroleum definitions and terminology that have not been covered and require further mention here because of the direct relationship to the pricing of petroleum and petroleum products.

Terminology is the means by which various subjects are named so that reference can be made in conversations and in writings and so that the meaning is passed on. Definitions are the means by which scientists and engineers communicate the nature of a material to each other and to the world, through either the spoken or the written word. Thus, the definition of a material can be extremely important and have a profound influence on how the technical community and the public perceive that material. In fact the definition of petroleum has been varied, unsystematic, diverse, and often archaic. Furthermore, the terminology of petroleum is a product of many years of growth. The long established use of an expression, however inadequate it may be, is altered with difficulty, and a new term is at best adopted slowly.

In general, the large reservoirs of light oil resources have been exploited first because larger reservoirs are easier to find and exploit and lighter oils are more valuable and require less energy to extract and refine to desirable products. Therefore, over time in mature regions, lower quality crude oil (often called heavy crude oil) has often required the exploitation of increasingly small, deep, and heavy offshore resources. Progressive depletion also means that oil in older fields that once came to the surface through natural drive mechanisms, such as gas pressure, must now be extracted using energy-intensive secondary and enhanced technologies. Another aspect of the quality of an oil resource is that oil reserves are normally defined by their degree of certainty and their ease of extraction, classed as proven, probable, possible or speculative. In addition, there are unconventional resources such as heavy oil, deep-water oil, tar sand bitumen (oil sand bitumen in Canada) and shale oil that are very energy intensive and costly to exploit.

However, the true reserves of a field will be known absolutely only on the day when it is finally abandoned — when cumulative production is know with the highest degree of certainty and subject only to counting errors. Prior to that, reserves are known only within a spectrum of uncertainty that should be expressed in terms of a probability range. The traditional method, which classifies reserves as proven, probable, or possible, tends to ignore the range of uncertainty, giving a single number for each class of reserves. The Securities and Exchange Commission only accepts the proven class for financial reporting purposes.

Petroleum is found in various countries (Table 1.3) and is scattered throughout the earth's crust, which is divided into natural

Table 1.3 Oil reserves by country (Billions of barrels, one billion = 1×10^9).

Country	Reserves* Barrels $\times 10^9$
Canada	179
Iran	126
Iraq	115
Kuwait	102
Russia	60
Saudi Arabia	262
United Arab Emirates	98
United States	21
Venezuela	79
Other	238

* Source: U.S. Energy information administration, 2006.

groups or strata, categorized in order of their antiquity (Speight, 2007). These divisions are recognized by the distinctive systems of organic debris (as well as fossils, minerals, and other characteristics), which form a chronological time chart that indicates the relative ages of the earth's strata. It is generally acknowledged that carbonaceous materials such as petroleum occur in all these geological strata from the Precambrian to the recent, and the origin of petroleum within these formations is a question that remains open to conjecture and the basis for much research.

Petroleum is by far the most commonly used source of energy, especially as the source of liquid fuels (Table 1.4). Indeed, because of the wide use of petroleum, the past 150 years could very easily be dubbed the Oil Century-and-a-Half, the Petroleum Era (c.f. the Pleistocene Era), or the New Rock Oil Age (c.f. the New Stone Age) (Ryan, 1998). For example, the United States imported approximately 6,000,000 barrels per day of petroleum and petroleum products in 1975 and now imports approximately double this amount. The majority of the products of a refinery are fuels (Pellegrino, 1998) and it is evident that this reliance on petroleum-based fuels

Table 1.4 Current and projected energy consumption scenarios (GRI, 1998).

	Energy Consumption (quads)					
	1995	**1996**	**2000**	**2005**	**2010**	**2015**
	Actual		**Projected**			
Petroleum	34.7	36.0	37.7	40.4	42.6	44.1
Gas	22.3	22.6	23.9	26.3	28.8	31.9
Coal	19.7	20.8	22.3	24.1	26.2	29.0
Nuclear	7.2	7.2	7.6	7.4	6.9	4.7
Hydro	3.4	4.0	3.1	3.2	3.2	3.2
Other	3.2	3.3	3.7	4.0	4.8	5.2

and products will continue for several decades. As a result, fossil fuels are projected to be the major sources of energy for the next fifty years. In this respect, petroleum and its associates (heavy oil and residua) are extremely important in any energy scenario, especially those scenarios that relate to the production of liquid fuels.

Indeed, over the past two decades the quality of crude oil has deteriorated (Swain, 1991, 1993, 1998, 2000), which has caused the nature of refining to change considerably. This, of course, has led to the need to manage crude quality more effectively through evaluation and product slates (Waguespack and Healey, 1998; Speight, 2007). Indeed, the declining reserves of lighter crude oil have resulted in an increasing need to develop options to desulfurize and upgrade the heavy feedstocks, specifically heavy oil and bitumen (Speight, 2008). This has resulted in a variety of process options that specialise in sulfur removal during refining. Though it will not be covered in this text, it is worthy of note that microbial desulfurization is being assiduously investigated as a recognised commercial technology for desulfurization (Monticello, 1995; Armstrong *et al.*, 1997).

With the necessity of processing heavy oil, bitumen, and residua to obtain more gasoline and other liquid fuels, there has been the recognition that knowledge of the constituents of these higher boiling feedstocks is also of some importance. Indeed, the problems encountered in processing the heavier feedstocks can be equated to the chemical character and the amount of complex, higher-boiling constituents in the feedstock. Refining these materials is not just a matter of applying know-how derived from refining conventional crude oils, but also requires knowledge of the chemical behavior of these more complex constituents. Furthermore, although the elemental analysis of tar sand bitumen has also been widely reported (Wallace *et al.*, 1988; Speight, 1990), the data often suffers from the disadvantage that identification of the source is too general and not always site specific.

With all of the scenarios in place, there is no doubt that petroleum and its relatives: residua, and heavy oil, as well as tar sand bitumen will be required to produce a considerable proportion of liquid fuels into the foreseeable future. Desulfurization processes will be necessary to remove sulfur in an environmentally acceptable manner to produce environmentally acceptable products. Refining strategies will focus on upgrading the heavy oils and residua and will emphasize the differences between the properties of the feedstocks. This will dictate the choice of methods or combinations thereof for the conversion of these materials to products (Schuetze and Hofmann, 1984).

Because of the need for a thorough understanding of petroleum and the associated technologies, it is essential that the definitions and terminology of petroleum science and technology be given prime consideration. This will aid in a better understanding of petroleum, its constituents, and its various fractions. Of the many forms of terminology that have been used, not all have survived, but the more commonly used are illustrated here. Particularly troublesome and more confusing are those terms that are applied to the more viscous materials, including the use of the terms *bitumen* and *asphalt*. This part of the text attempts to alleviate much of the confusion that exists, but it must be remembered that the terminology of petroleum is still open to personal choice and historical usage.

1.4.1 Petroleum

When petroleum occurs in a reservoir that allows the crude material to be recovered by pumping operations as a free-flowing dark to light colored liquid, it is often referred to as conventional petroleum.

Petroleum is a naturally occurring mixture of diverse hydrocarbons whose physical and chemical qualities reflect the different origins and, especially, different degrees of natural processing of these hydrocarbons (Speight, 2007, 2008). In fact, the term petroleum covers a wide assortment of materials consisting of mixtures of hydrocarbons and other compounds containing variable amounts of sulfur, nitrogen, and oxygen, which may vary widely in volatility, specific gravity, and viscosity. Metal-containing constituents, notably those compounds that contain vanadium and nickel, usually occur in the more viscous crude oils in amounts up to several thousand parts per million, and can have serious consequences during processing of these feedstocks (Speight, 2007). Because petroleum is a mixture of widely varying constituents and proportions, its physical properties also vary widely and the color from colorless to black.

There is also another type of petroleum that is different from the conventional petroleum insofar as they are much more difficult to recover from the subsurface reservoir. This material, or heavy oil, has a much higher viscosity and lower API gravity than conventional petroleum and recovery of heavy oil usually requires thermal stimulation of the reservoir (Speight, 2007; Speight, 2008).

The majority of crude oil reserves identified to date are located in a relatively small number of very large fields, known as giants. In fact, approximately 300 of the largest oil fields contain almost 75% of the available crude oil. Although most of the world's nations produce at least minor amounts of oil, the primary concentrations are in the Persian Gulf, North and West Africa, the North Sea, and the Gulf of Mexico. In addition, of the 90 oil-producing nations, five Middle Eastern countries contain almost 70% of the current, known oil reserves.

For many years, petroleum and heavy oil were very generally defined in terms of physical properties. For example, heavy oils were considered to be those crude oils that had gravity somewhat less than 20°API with the heavy oils falling into the API gravity range 10 to 15°. For example, Cold Lake heavy crude oil has an API gravity equal to 12°, and tar sand bitumen usually has an API gravity in the range 5–10° (Athabasca bitumen = 8°API). Residua would vary depending upon the temperature at which distillation was terminated, but usually vacuum residua fall into the approximate range 2 to 8°API. However, the classification of crude oil by the use of a single physical property is subject to the errors inherent in the analytical method (by which the property is determined) and must be used with caution (Chapter 4).

1.4.2 Natural Gas

The generic term natural gas applies to gases commonly associated with petroliferous (petroleum-producing, petroleum-containing) geologic formations. Natural gas generally contains high proportions of methane (a single carbon hydrocarbon compound, CH_4) and some of the higher molecular weight higher paraffins (C_nH_{2n+2}) generally containing up to six carbon atoms may also be present in small quantities (Table 1.5). The hydrocarbon constituents of natural gas are combustible, but non-flammable non-hydrocarbon components such as carbon dioxide, nitrogen, and helium are often present in the minority and are regarded as contaminants.

In addition to the natural gas fund in petroleum reservoirs, there are also those reservoirs in which natural gas may be the sole occupant. The principal constituent of natural gas is methane, but other hydrocarbons, such as ethane, propane, and butane, may also be present. Carbon dioxide is also a common constituent of natural gas. Trace amounts of rare gases, such as helium, may also occur, and certain natural gas reservoirs are a source of these rare gases. Just as petroleum can vary in composition, so can natural gas. Differences in natural gas composition occur between different reservoirs, and

Table 1.5 Constituents of natural gas.

Name	Formula	Vol. %
Methane	CH_4	>85
Ethane	C_2H_6	3–8
Propane	C_3H_8	1–5
Butane	C_4H_{10}	1–2
Pentane$^+$	C_5H_{12}	1–5
Carbon dioxide	CO_2	1–2
Hydrogen sulfide	H_2S	1–2
Nitrogen	N_2	1–5
Helium	He	<0.5

Pentane$^+$: pentane and higher molecular weight hydrocarbons, including benzene and toluene.

two wells in the same field may also yield gaseous products that are different in composition (Speight, 1990).

Natural gas (also called *marsh gas* and *swamp gas* in older texts and more recently *landfill gas*) is a gaseous fossil fuel that is found in oil reservoirs, natural gas reservoirs, coal seams. It is a vital component of the world's supply of energy. It is one of the cleanest, safest, and most useful of all energy sources. While it is commonly grouped in with other fossil fuels and sources of energy, there are many characteristics of natural gas that make it unique.

Natural gas is the result of the decay of animal remains and plant remains, or organic debris, that has occurred over millions of years. Over time, the mud and soil that covered the organic debris changed to rock and trapped the debris beneath the newly formed rock sediments. Pressure, and to some extent, heat (as yet undefined) changed some of the organic material into coal, some into oil (petroleum), and some into natural gas. Whether or not the debris formed coal, petroleum, or gas depended upon the nature of the debris and the localized conditions under which the changes occurred.

Natural gas has been known for many centuries, but its initial use was probably more for religious purposes rather than as a fuel. For example, gas wells were an important aspect of religious life in ancient Persia because of the importance of fire in their religion. In classical times, these wells were often flared and must have been awe-inspiring (Scheil and Gauthier, 1909; Schroder, 1920; Lockhart, 1939; Forbes, 1964). Furthermore, just as petroleum was used in antiquity, natural gas was also known in antiquity. However, the use of petroleum has been relatively well documented because of its use in warfare and as mastic for walls and roads (Henry, 1873; Abraham, 1945; Forbes, 1958a, 1958b; James and Thorpe, 1994).

Natural gas was first discovered in the United States in Fredonia, New York, in 1821. In the years following this discovery, natural gas usage was restricted to its environs because the technology for storage and transportation (bamboo pipes notwithstanding) was not well developed and, at that time, natural gas had little or no commercial value. In fact, in the 1930s when petroleum refining was commencing — an expansion in technology that is still continuing — natural gas was not considered a major fuel source and was only produced as an unwanted by-product of crude oil production and/or refining. It is only during the last several decades that natural gas has been seen as a major contributor to energy production.

There are several general definitions that have been applied to natural gas. For example, associated or dissolved natural gas occurs either as free gas or as gas in solution in the petroleum. Gas that occurs as a solution in the petroleum is dissolved gas, whereas the gas that exists in contact with the petroleum (gas cap) is associated gas. In addition, lean gas is gas in which methane is the major constituent and wet gas contains considerable amounts of the higher molecular weight hydrocarbons. Sour gas contains hydrogen sulfide, whereas sweet gas contains very little, if any, hydrogen sulfide. In direct contrast to the terminology of the petroleum industry where the residue (residuum, resid) is the high boiling material left after distillation, residue gas is natural gas from which the higher molecular weight (higher boiling) hydrocarbons have been extracted and so is the lowest boiling hydrocarbon in natural gas. Finally casing head gas (casinghead gas) is derived from petroleum, but is separated at the well-head separation facility.

To further define the terms dry and wet in quantitative measures, the term dry natural gas indicates that there is less than 0.1 gallon (1 gallon, US, = 264.2 m^3) of gasoline vapor (higher molecular weight paraffins) per 1000 ft^3 (1 ft^3 = 0.028 m^3). The term wet natural gas indicates that there are such paraffins present in the gas — in fact, more than 0.1 gal/1000 ft^3.

Other components such as carbon dioxide (CO_2), hydrogen sulfide (H_2S), mercaptans (thiols; R-SH), as well as trace amounts of other constituents may also be present. Thus, there is no single organization of components that maybe termed typical natural gas.

1.4.3 Heavy Oil

Heavy oil is a type of petroleum that is different from conventional petroleum insofar as it is much more difficult to recover from the subsurface reservoir and has a much higher viscosity (and lower API gravity) than conventional petroleum, and primary recovery of heavy oil usually requires thermal stimulation of the reservoir.

Petroleum and heavy oil have been very generally defined in terms of physical properties (Section 1.2). For example, heavy oil was considered to be crude oil that had gravity somewhat less than 20°API with tar sand bitumen falling into the API gravity range <10°. For example, Cold Lake heavy crude oil has an API gravity equal to 12° and tar sand bitumen, usually has an API gravity in the range of 5–10° (Athabasca bitumen = 8°API). However, classification of crude oil by the use of a single physical property is subject to the errors inherent in the analytical method by which the property is determined and must be used with caution (Chapter 4).

While conventional crude oil flows naturally and can be pumped without being heated or diluted, heavy crude oil usually requires thermal stimulation to cause recovery. In fact, a more appropriate definition of heavy oil is that it is it is recoverable in its natural state by conventional oil well production methods including currently used enhanced recovery techniques. By analogy, tar sand bitumen it is not recoverable in its natural state using enhanced (tertiary) recovery techniques (Section 5.4).

The term extra heavy oil has been introduced fairly recently without reasonable justification and often serves to confuse the real issues of nomenclature. It is used to define the material (such as tar sand bitumen) that occurs in the near-solid state and is incapable of free flow under ambient conditions. Yet a definition of such material has been available for four decades (Section 1.5) and use of the term extra heavy oil only serves to confuse matters even further by introducing another unknown in the simple arena of terminology and definitions.

1.4.4 Tar Sand Bitumen

Throughout this text, frequent reference is made to tar sand bitumen, but because commercial operations have been in place for over 40 years (Spragins, 1978; Speight, 1990) it is not surprising that more is known about the Alberta, Canada tar sand reserves than any other reserves in the world. Therefore, when discussion is made of tar sand deposits, reference is made to the relevant deposit, but when the information is not available, the Alberta material is used for the purposes of the discussion.

The term bitumen (also, on occasion, referred to as native asphalt) includes a wide variety of naturally occurring reddish brown to black materials of semisolid, viscous to brittle character that can exist in nature with no mineral impurity or with mineral matter contents that exceed 50% by weight. Bitumen is frequently found filling pores and crevices of sandstone, limestone, or argillaceous sediments, in which case the organic and associated mineral matrix is known as rock asphalt (Abraham, 1945; Hoiberg, 1964). Tar sand bitumen is a high-boiling material with little, if any, material boiling below 350°C (660°F). The term oil sand is also used in the same way as the term tar sand, and these terms are used interchangeably throughout this text.

Bitumen is also a common term used in many European countries for the mastic that makes up road asphalt — a mixture of asphalt and aggregate. The term bitumen is not used in any such sense in this book.

Furthermore, it is incorrect to refer to naturally occurring bitumen as tar or pitch. Although the word tar is somewhat descriptive of the black bituminous material, it is best to avoid its use with respect to natural materials. More correctly, the name tar is usually applied to the heavy product remaining after the destructive distillation of coal or other organic matter. Pitch is the distillation residue of the various types of tar. Alternative names, such as bituminous sand or oil sand, are gradually finding usage, with the former name more technically correct.

For the purposes of this text, the definition of tar sand bitumen is derived from the definition of tar sand that has been defined by the United States government (FE-76-4):

> [Tar sands]...the several rock types that contain an extremely viscous hydrocarbon which is not recoverable in its natural state by conventional oil well production methods including currently used enhanced recovery techniques. The hydrocarbon-bearing rocks are variously known as bitumen-rocks oil, impregnated rocks, oil sands, and rock asphalt.

By inference, heavy oil (Section 3.1) is a resource that can be recovered in its natural state by conventional oil well production methods including currently used enhanced recovery techniques. The term natural state means without conversion of the heavy oil or bitumen as might occur during thermal recovery processes.

Because of the diversity of available information and the continuing attempts to delineate the various world tar sand deposits, it is virtually impossible to present accurate numbers that reflect the extent of the reserves in terms of the barrel unit. The term extra heavy oil (Section 1.4) is often invoked without explanation or adequate description, leaving the reserves of heavy oil and tar sand bitumen open to speculation. Indeed, investigations into the extent of many of the world's deposits are continuing at such a rate that the numbers vary from one year to the next. Accordingly, the data quoted here must be recognized as approximate with the potential of being quite different at the time of publication.

Nevertheless, whatever numbers are used, bitumen in tar sand deposits represent a potentially large supply of energy (Chapter 2). However, many of these reserves are only available with some difficulty and optional refinery scenarios will be necessary for conversion of these materials to low-sulfur liquid products because of the substantial differences in character between conventional petroleum and tar sand bitumen. Bitumen recovery requires the prior application of reservoir fracturing procedures before the introduction of thermal recovery methods. Currently, commercial operations in Canada use mining techniques for bitumen recovery.

Even though tar sand deposits are widely distributed throughout the world (Speight, 1990, 2007), the fact that commercialization has taken place in Canada does not mean that commercialization is imminent for other tar sands deposits. There are considerable differences between the Canadian and the US deposits that could preclude across-the-board application of the Canadian principles to the US sands (Speight, 1990). Whilst Canadian scientist and engineers know much about the Athabasca (Alberta) deposits, the knowledge is not generally applicable to other deposits — the key is site specificity in terms of accessibility and recoverability.

1.5 References

Abraham, H. 1945. Asphalts and Allied Substances. Van Nostrand, New York.

Armstrong, S.M., Sankey, B.M., and Voordouw, G. 1997. Fuel. 76: 223.

Barton, G.A. 1926. On Binding-reeds, Bitumen, and Other Commodities in Ancient Babylonia. J. Amer. Orient. Soc. 46: 297–302.

Bauer, G. (Georgius Agricola) (1546). Book IV, De Natura Fossilium, Basel, Switzerland.

Bauer, G. (Georgius Agricola) (1556). Book XII, De Re Metallica, Basel.

Bell, H. S. (1945). American Petroleum Refining, van Nostrand, New York.

Boyle, G. (Editor). 1996. Renewable Energy: Power for a Sustainable Future. Oxford University Press, Oxford, England.

Cobb, C., and Goldwhite, H. 995. Creations of Fire: Chemistry's Lively History from Alchemy to the Atomic Age. Plenum Press, New York.

Cook, A. B., and Despard, C. (1927). J. Inst. Petroleum Technol. 13:124.

Dahmus, J. 1995. A History of the Middle Ages. Barnes and Noble, New York. P. 135 and 136; previously published with the same title by Doubleday Book Co., 1968.

Davies, N. 1996. Europe: A History. Oxford University Press. Oxford, England. p. 245 and 250.

Forbes, R. J. 1958a. A History of Technology, Oxford University Press, Oxford, England.

Forbes, R. J. 1958b. Studies in Early Petroleum Chemistry. E. J. Brill, Leiden, The Netherlands.

Forbes, R.J. 1959. More Studies in Early Petroleum Chemistry. E.J. Brill, Leiden, The Netherlands:

Forbes, R. J. 1964. Studies in Ancient Technology. E. J. Brill, Leiden, The Netherlands.

GRI, 1998. Baseline. Report No. GRI-98/0001. Gas Research Institute, Chicago, IL. March. p. 1.

Henry, J.T. 1873. The Early and Later History of Petroleum. Volumes I and II. APRP Co., Philadelphia, PA.

Hoiberg, A. J. 1964. Bituminous Materials: Asphalts, Tars, and Pitches. John Wiley & Sons. New York.

James, P., and Thorpe, N. 1994. Ancient Inventions. New York: Ballantine Books.

Johnson, P. 1997. Harper Collins Publishers Inc., New York. p. 602

Kovacs, M.G. 1990. The Epic of Gilgamesh. Stanford: Stanford University Press, 1990). Tablet XI.

Krishnan, J.M., and Rajagopal, K.R. 2003. Review of the Uses and Modeling of Bitumen from Ancient to Modern Times. Appl. Mech., Rev. 56(2): 149–205.

Lockhart, L. 1939. J. Inst. Petroleum 25: 1.

Mallowan, M.E.L. (1954). Iraq 16(2):115.

Mallowan, M.E.L., and Rose, J. C. 1935. Iraq 2(1):1.

Marschner, R.F., Duffy, L.J., and Wright, H. (1978). Paleorient 4:97.

Monticello, D.J. 1995. United States Patent 5,387, 523. February 7.

Moorey, P.R.S. 1994. Ancient Mesopotamian Materials and Industries: The Archaeological Evidence, Clarendon Press, Oxford, England.

Nellensteyn, F. J., and Brand, J. (1936). Chem. Weekbl. 261.

Owen, E.W. 1975. In A History of Exploration for Petroleum. Memoir No. 6. American Association of Petroleum Geologists, Tulsa, Oklahoma, USA. P. 2.

Pellegrino, J.L. 1998. Energy and Environmental Profile of the US Petroleum: Refining Industry. Office of Industrial Technologies. United States Department of Energy, Washington, DC.

Pirog, R. 2007. The Role of National Oil Companies in the International Oil Market August 21, 2007. Report No. RL34137. CRS Report for Congress, Congressional Research Service, Washington, DC. August 21.

Powicke, Sir Maurice. 1962. The Thirteenth Century. 1216–1307. 2nd Edition. Oxford University Press, Oxford, England. p. 13.

Ramage, J. 1997. Energy: A Guidebook. Oxford University Press, Oxford, England.

Ryan, J.F. 1998. Today's Chemist at Work. 7(6): 84.

Schuetze, B., and Hofmann, H. 1984. Hydrocarbon Processing. 63(2): 75.

Scheil, V., and Gauthier, A. 1909. Annales de Tukulti Ninip II, Paris.

Schroder, O. 1920. Keilschriftetexte aus Assur Vershiedenen, xiv. Leipzig.

Speight, J.G. 1978. Personal observations at archeological digs at the sites of the ancient Iraqi cities of Babylon, Calah, Nineveh, and Ur.

Speight, J. G. 1990. In Fuel Science and Technology Handbook. J.G. Speight (Editor). Marcel Dekker, New York. Chapters 12–16.

Speight, J.G. 2007. The Chemistry and Technology of Petroleum. 4th Edition. CRC Taylor & Francis Group, Boca Raton, Florida.

Speight, J.G. 2008. Handbook of Synthetic Fuels, McGraw-Hill, New York.

Sperber, D. 1976. Objects of Trade Between Palestine and Egypt in Roman Times. Journal of the Economic and Social History of the Orient. 19(2): 113–147.

Spragins, F.K. 1978. Development in Petroleum Science, No. 7, Bitumens, Asphalts and Tar Sands. T.F. Yen and G.V. Chilingarian (Editors). Elsevier, New York, p. 92.

Steinsaltz, 1977. The Uses and Refining of Petroleum as Mentioned in the Talmud. Isis, 68(1): 104–105.

Stern, B. Connan, J., Blakelock, E., Jackman, R., Coningham, R.A.E., and Heron, C. 2007. From Susa to Anuradhapura: Reconstructing Aspects of Trade and Exchange in Bitumen-Coated Ceramic Vessels between Iran and Sri Lanka from the Third to the Ninth Centuries AD. Archaeometry, 50(3): 409–428.

Swain, E.J. 1991. Oil & Gas Journal. 89(36): 59.

Swain, E.J. 1993. Oil & Gas Journal. 91(9): 62.

Swain, E.J. 1998. Oil & Gas Journal. 96(40): 43.

Swain, E.J. 2000. Oil & Gas Journal. March 13.

Thompson, R.C. 1936. Dictionary of Assyrian Chemistry and Geology, Clarendon Press, Oxford, England.

Wendt, C.J., and Lu, S-T. 2006. Sourcing Archeological Bitumen in the Olmec Region. Journal of Archeological Science, 33(1): 89–97.

Wendt, C.J., and Cyphers, A. 2008. How the Olmec used Bitumen in Ancient Mesoamerica. Journal of Anthropological Archeology, 27(2): 172–191.

Waguespack, K.G., and Healey, J.F. 1998. Hydrocarbon Processing. 77(9): 133.

Wallace, D., Starr, J., Thomas, K.P., and Dorrence. S.M. 1988. Characterization of Oil Sands Resources. Alberta Oil Sands Technology and Research Authority, Edmonton, Alberta, Canada.

2

Origin and Occurrence of Oil

Petroleum has been used as an energy resource for more than 6,000 years (Chapter 1) and it is by far the most commonly used source of energy, especially as the source of liquid fuels and use of petroleum is projected to continue at least in current amounts for at least two decades (Speight, 2007a; BP, 2008; World Energy Council, 2008; Speight, 2011) (Figure 2.1). In recent years, the average quality of crude oil has become worse. This is reflected in a progressive decrease in API gravity (i.e., increase in density) and a rise in sulfur content (Swain, 1991, 1993, 1997, 2000). It is now believed that there has been a recent tendency for the quality of crude oil feedstocks to stabilise. Be that as it may, the nature of crude oil refining has been changed considerably.

The declining reserves of light crude oil have resulted in an increasing need to develop options to upgrade the abundant supply of known heavy oil reserves. In addition, there is considerable focus and renewed efforts on adapting recovery techniques to the production of heavy oil and tar sand bitumen (Speight, 2007a).

Fossil fuels are those fuels, namely coal, petroleum (including heavy oil and bitumen), natural gas, and oil shale produced by the

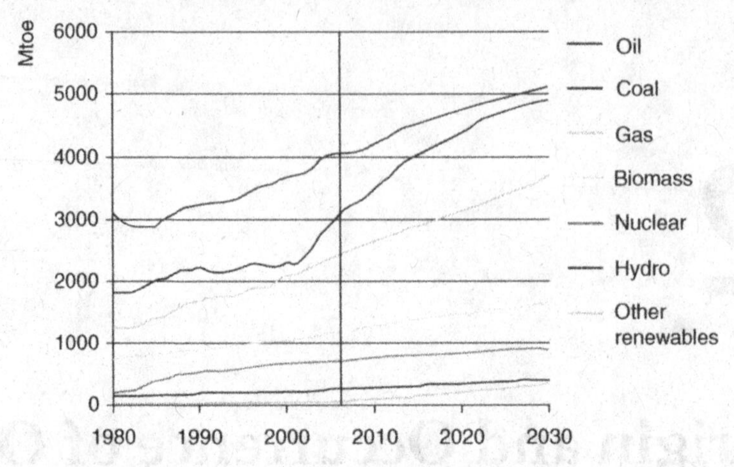

Figure 2.1 *Distribution of World Energy Resources, millions of tons oil equivalent (Mtoe).
*Source: World Energy Outlook 2008, International Energy Agency.

decay of plant remains over geological time (Speight, 1990, 2008). Resources such as heavy oil and bitumen in tar sand formations, which are also discussed in this text, represent an unrealized potential (Speight, 2008, 2009), with liquid fuels from petroleum being only a fraction of those that could ultimately be produced from heavy oil and tar sand bitumen.

In fact, at the present time, the majority of the energy consumed by human society is produced from the fossil fuels (petroleum: ca. 38 to 40%, coal: ca. 31 to 35%, natural gas: ca. 20 to 25%) with the remainder of the energy requirements to come from nuclear and hydroelectric sources. As a result, fossil fuels (in varying amounts depending upon the source of information) are projected to be the major sources of energy for the next fifty years.

Petroleum is scattered throughout the earth's crust, which is divided into natural groups or strata, categorized in order of their antiquity (Table 2.1). These divisions are recognized by the distinctive systems of organic debris (as well as fossils, minerals, and other characteristics) that form a chronological time chart that indicates the relative ages of the earth's strata. It is generally acknowledged that carbonaceous materials such as petroleum occur in all these geological strata from the Precambrian to the recent, and the origin of petroleum within these formations is a question that remains open to conjecture and the basis for further investigation.

Table 2.1 The geologic timescale.

Era	Period	Epoch	Approximate Duration (Millions of Years)	Approximate Number of Years Ago (Millions of Years)
Cenozoic	Quaternary	Holocene	10,000 years ago to the present	
		Pleistocene	2	.01
	Tertiary	Pliocene	11	2
		Miocene	12	13
		Oligocene	11	25
		Eocene	22	36
		Paleocene	71	58
Mesozoic	Cretaceous		71	65
	Jurassic		54	136
	Triassic		35	190

Table 2.1 (cont.) The geologic timescale.

Era	Period	Epoch	Approximate Duration (Millions of Years)	Approximate Number of Years Ago (Millions of Years)
Paleozoic	Permian		55	225
	Carboniferous		65	280
	Devonian		60	345
	Silurian		20	405
	Ordovician		75	425
	Cambrian		100	500
Precambrian			3,380	600

2.1 The Formation of Oil

There are two theories on the origin of petroleum: the abiogenic theory and the biogenic theory (Kenney *et al.*, 2001, 2002). Both theories have been intensely debated since the 1860s, shortly after the discovery of widespread occurrence of petroleum.

The abiogenic theory of petroleum formation postulates that inorganic substances were the source material. In fact, there are several recent theories related to the formation of petroleum from non-biogenic sources in the earth (Gold, 1984, 1985; Gold and Soter, 1980, 1982, 1986; Osborne, 1986; Szatmari, 1989).

Thus, the idea of abiogenic petroleum origin proposes that large amounts of carbon exist naturally in the planet, some in the form of hydrocarbons. Hydrocarbons are less dense than aqueous pore fluids, and migrate upward through deep fracture networks. Thermophilic, rock-dwelling microbial life forms are in part responsible for the biomarkers found in petroleum. However, their role in the formation, alteration, or contamination of the various hydrocarbon deposits is not yet understood. Thermodynamic calculations and experimental studies confirm that n-alkanes (common petroleum components) do not spontaneously evolve from methane at pressures typically found in sedimentary basins, and so the theory of an abiogenic origin of hydrocarbons suggests deep generation below 200 km.

However, most geologists and geochemists support the biogenic theory of petroleum formation and consider crude oil and natural gas as the products of compression and heating of ancient vegetation over geological time, although the chemistry of the transformation of the organic matter into petroleum is not well understood and is largely speculative (Tissot and Welte, 1978; Snowdon and Powell, 1982; Brooks and Welte, 1984; Speight, 2007a). Over millennia, this organic matter — the decayed remains of prehistoric marine animals and terrestrial plants — mixed with mud and was buried under thick sedimentary layers of material. While the minimum and maximum temperatures have never been fully defined or conclusively proven, heat and pressure caused the remains to metamorphose, first into a material known as protopetroleum, and then into liquid and gaseous hydrocarbons by means of a maturation process (catagenesis) (Califet and Oudin, 1966; Barker and Wang, 1988; Speight, 2007a). The liquid and gaseous hydrocarbons then migrated through adjacent rock layers until they were trapped underground in porous rock formations, or reservoirs. Hydrocarbon accumulations in a reservoir require that the reservoir rock has the capacity of transmit fluids (permeability) and to store fluids (porosity) as well as an impermeable basement rock and an impermeable cap rock to prevent their escape.

The arguments against the abiotic theory involve various biomarkers which have been found in all samples of all the oil and gas accumulations found to date. The prevailing view among geologists and petroleum engineers is that this evidence provides irrefutable proof that the oil and gas accumulations found up to now have a biologic origin.

The composition of the precursors has a major influence on the composition of petroleum and the relative amounts of these precursors (dependent upon the local flora and fauna) that occurred in the source material added another variable to the composition of the produced oil. Hence, it is not surprising that petroleum composition can vary with the location and age of the field, in addition to any variations that occur with the depth of the individual well. Two adjacent wells are more than likely to produce petroleum with very different characteristics.

2.2 Reservoirs

Reservoir location and structure plays an important role in crude oil economics — some would even say the most important role.

A reservoir is any porous and permeable stratum that will store crude oil and a very common reservoir rock is a porous or fractured limestone (especially of the reef, bioherm type); several such reservoir structures are found throughout the earth. Most reservoir rocks are sedimentary rocks, almost always the coarser grained of the sedimentary rocks: sand, sandstone, limestone, and dolomite. A less common reservoir is a fractured shale or even igneous or metamorphic rock. It is only rarely that shale acts as a reservoir rock, and again fractures and other relatively wide openings are believed to confer the required reservoir properties on an otherwise unsuitable rock.

2.2.1 Reservoir Structure

Petroleum accumulations are usually found in structural highs (an anticline), where reservoir rocks of suitable porosity and permeability are covered by a dense, relatively impermeable cap rock, such as an evaporite or shale. A reservoir rock sealed by a cap rock in the position of such a geological high (i.e., the anticline) is known as a structural petroleum trap (Figure 2.2). Stratigraphic traps are also known and occur in various fields. In all cases, changes in permeability and porosity determine the location of oil and/or a gas accumulation. Such accumulations may be several miles in length.

Structural traps are formed by a deformation in the rock layer that contains the hydrocarbons. Anticlines, domes, and folds are common structures. Fault-related features also may be classified as structural traps if closure is present. Structural traps are the easiest to locate by surface and subsurface geological and geophysical studies. They are the most numerous among traps and have received a greater amount of attention in the search for oil than all other types of traps. On the other hand, stratigraphic traps are formed when other beds seal a reservoir bed or when the facies change (change in the rock-type) within the reservoir bed itself leading to a change in permeability and stoppage of the flow of the fluids.

The simplest form of the structural trap is the anticline and the dome, each of which has a convex upper surface (Figure 2.2). Many oil and gas accumulations are trapped in anticlines or domes, structures that are generally more easily detected than some other types of traps. There are also examples in which the reservoir rock wedges out at its upper end as an original depositional feature due to lateral variation in deposition or abuts against an old land surface (stratigraphic trap). Traps associated with salt intrusions are

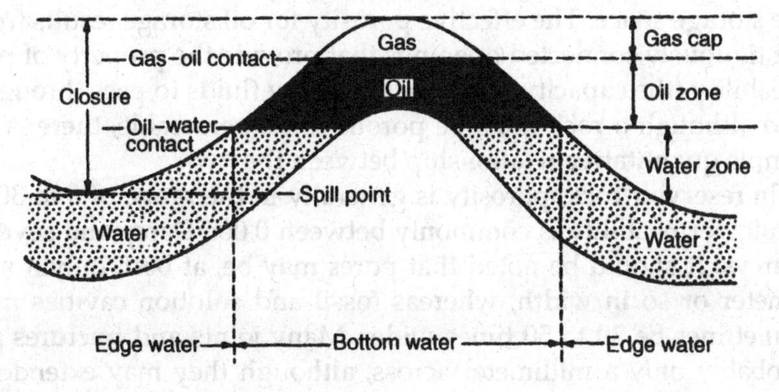

Figure 2.2 Typical anticlinal petroleum trap.

of various kinds; limestone reefs can also serve as reservoir rocks and give rise to overlying traps of anticlinal form as a result of differential compaction. Lastly, examples are also known in which the reservoir rock extends to the surface of the earth but oil and gas are sealed in it by clogging of the pores by bitumen or by natural cements. Many reservoirs display more than one of the factors contributing to the entrapment of hydrocarbons.

The distinction between a structural trap and a stratigraphic trap is often difficult to define. For example, an anticlinal trap may be related to an underlying buried limestone reef. Beds of sandstone may wedge out against an anticline because of depositional variations or intermittent erosion intervals. Salt domes, formed by flow of salt at substantial depths, also have created numerous traps that are both a structural trap and a stratigraphic trap.

2.2.2 Accumulation in Reservoirs

To all intents and purposes and to the casual observer, a rock is solid. Yet a rock can possess the fluid holding capacity by virtue of the uneven contact between the mineral grains and the ensuing space (porosity) within the rock.

Reservoir rock (usually, but not always, sandstone) must possess fluid-holding capacity (porosity) and also fluid-transmitting capacity (permeability). A variety of different types of openings in rocks are responsible for these properties in reservoir rocks, the most common of which are the pores between the grains of which the rocks are made or the cavities inside fossils, openings formed by solution or fractures, and joints that have been created in various ways. The relative proportion of the different kinds of openings varies with the rock type, but pores usually account for the bulk of the storage space. The effective porosity for oil storage results from continuously connected openings that provide the property of permeability (the capacity of the rock to allow fluids to pass through), and although a rock must be porous to be permeable, there is no simple quantitative relationship between the two.

In reservoir rock, porosity is generally in the range of 5 to 30%; while permeability is commonly between 0.005 Darcy and several Darcys. It should be noted that pores may be, at best, only a millimeter or so in width, whereas fossil and solution cavities may sometimes be 30 to 50 times wider. Many joints and fractures are probably only a millimeter across, although they may extend for considerable distances.

Petroleum accumulations are generally found in relatively coarse-grained porous and permeable rocks that contain little or no insoluble organic matter. It is highly unlikely that petroleum originated from organic matter of which no trace remains now. Rather, it appears that petroleum constituents are generated through geochemical action on organic detritus that is usually found in fine-grained sedimentary rocks. In addition, it may be anticipated that some of the organic material would also change sufficiently from the original organic material to remain in the sediment through and beyond the oil-generating stage. Thus, it is generally assumed that the place of origin of oil and gas is not identical to the locations where it is found. The oil has had to migrate to present reservoirs from the place of origin.

When sediments such as mud and clay are deposited, they contain water. As the layers of sediment accumulate, the increasing load of material causes compaction of the sediment and part of the water is expelled from the pores. *Migration* of oil probably begins at this stage and may involve a substantial horizontal component. Initially, the oil hydrocarbons may be present in the water, either suspended as tiny globules or dissolved, with hydrocarbons being slightly soluble in water. The loss of hydrocarbons from a trap is referred to as dismigration.

However, the processes by which petroleum migrates into pools are not well defined and are speculative (Table 2.2), but it is known that crude oil and gas occur in pools or fields in which the oil or gas occupies pore space in the rock. Some pools are large, extending laterally over many square kilometers, with their vertical extent ranging from several meters to as much as several hundred meters. The crude oil or gas occupying the pore space in the pool has displaced the water that was initially present in the pores. The presence of a pool of oil or gas therefore implies that the oil or gas has migrated into the pool. Within the pool, there may be a single phase (oil and gas dissolved in water) or a multiphase (separate water and hydrocarbon phases) fluid system.

2.2.3 Distribution of Fluids in the Reservoir

The specific gravity of gas and petroleum, the latter generally between specific gravity (at 60° F, 15.6°C) that varies from about 0.75 to 1.00 (57 to 10° API), with the specific gravity of most crude oils falling in the range 0.80 to 0.95 (45 to 17°API), are considerably lower than those of saline pore waters (specific gravity: 1.0 to 1.2).

Table 2.2 General mechanisms for petroleum migration.

Geological Event	Migration Effect
Basin development	downward fluid flow
Mature basin	sediments move downward
Hydrocarbon generation	thermal effects on source sediment
Hydrocarbons dissolve	pore fluid become saturated
Geothermal gradient changes	isotherms depressed
Pore fluids cool	hydrocarbons form separate phase
Hydrocarbons separate	move to top of carrier fluid (water)
Updip migration	buoyancy effect
Intermittent faulting	hydrocarbon migration to traps

The distribution of the fluids in a reservoir rock is dependent on the densities of the fluids as well as on the properties of the rock. If the pores are of uniform size and evenly distributed, there is:

1. an upper zone where the pores are filled mainly by gas (the gas cap).
2. a middle zone in which the pores are occupied principally by oil with gas in solution.
3. a lower zone with its pores filled by water.

Approximately 10 to 30% of water occurs along with the oil in the middle zone. There is a transition zone from the pores occupied entirely by water to pores occupied mainly by oil in the reservoir rock, and the thickness of this zone depends on the densities and interfacial tension of the oil and water, as well as the sizes of the pores. Similarly, there is some water in the pores in the upper gas zone that has at its base a transition zone from pores occupied largely by gas to pores filled mainly by oil.

The water found in the oil and gas zones is known generally as interstitial water. It usually occurs as collars around grain contacts, as a filling of pores with unusually small throats connecting with adjacent pores, or, to a much smaller extent, as wetting films on the surface of the mineral grains when the rock is preferentially wet by water. The water may occur as wetting films, or collars, around the sand grains as well as in some completely filled pores.

2.2.4 Migration of Reservoir Fluids

The predominant theory assumes that as the sedimentary layers superimposed in the source bed became thicker, the pressure increased and the compression of the source bed caused liquid organic matter to migrate to sediments with a higher permeability, which as a rule is sandstone (or porous limestone).

Primary migration is the movement of hydrocarbons from mature, organic-rich source rocks to a point where the oil and gas can collect as droplets or as a continuous phase of liquid hydrocarbon. Secondary migration is often attributed to various aspects of buoyancy and hydrodynamics (Schowalter, 1979; Barker, 1980) and is the movement of the hydrocarbons as a single, continuous fluid phase through water-saturated rocks, fractures, or faults followed by accumulation of the oil and gas in sediments (traps) from which further migration is prevented.

It is believed that during migration petroleum does not move through the bulk of the non-source rock and shale bodies but through faults and fractures that may be in the form of a channel network that permits leakage of oil and gas from one zone to another. During this migration, the composition of the oil may be changed through physical causes, such as filtration and adsorption, in a manner analogous to the chromatographic separation of petroleum. Reactions with minerals such as elemental sulfur or even with sulfur-containing minerals (e.g. sulfates) may also occur.

Source beds do not generally coincide with reservoir rocks and the belief is that petroleum, before it comes to rest in a trap, migrates large distances. Various examples are known of vertical migration, and migration upward for some kilometers is considered possible. Nevertheless the converse may also be true, and the in situ theory (juxtaposition of source bed and reservoir rock) advocates that petroleum migrates very little, if at all, and that even the amount of vertical migration is negligible.

Once the oil has accumulated in the reservoir rock, gravitational forces are presumed to be dominant, thereby causing the oil, gas, and water to segregate according to their relative densities in the upper parts of the reservoir (Landes, 1959). If the pores in the reservoir rock are of uniform size and evenly distributed, there are transition zones, from the pores occupied entirely by water to pores occupied mainly by oil to those pores occupied mainly by gas. The thickness of the water-oil transition zone depends on the densities and interfacial tension of the oil and water as well as on the size of the pores. Similarly, there is some water in the pores in the upper gas zone, which has at its base a transition zone from pores occupied largely by gas to pores filled mainly by oil.

The cap rock and basement rock, which are generally impermeable (or have a much lower permeability than the reservoir rock to oil and gas), act as a seal to prevent the escape of oil and gas from the reservoir rock. Typical cap and basement rocks are clay and shale, that is, strata in which the pores are much finer than those of reservoir rocks. Other rocks, such as marl and dense limestone, can also serve as cap and basement rocks provided that any pores are very small. There are cases in which evaporites (salt, anhydrite, and gypsum) act as effective sealants. The cap rock has a far lower permeability than the reservoir rock, but it is equally true that cap rocks have very high capillary pressures while those of reservoir rocks are much lower. The capillary pressure is the pressure required to cause a fluid to displace from the openings in a rock by another fluid with which it is not miscible. Capillary pressure is dependent on the size of the openings, the interfacial tension between the two fluids, and the contact angle for the system.

2.2.5 Transformation of Petroleum in the Reservoir

Petroleum is susceptible to alteration even after it has collected in a reservoir or in sediments (Evans *et al.*, 1971). The alteration process is ongoing and such alteration can change the quality of the oil during the life of the reservoir, thereby affecting the price of the oil. Therefore, it is important to briefly address the issue of the alteration of petroleum once it has accumulated in the reservoir.

Alteration of reservoir petroleum is accepted for most of the world oil accumulations and may be related to the relative instability of petroleum, or the traps may be susceptible to incursion to chemical agents, such as oxygen. Physical effects, such as those caused when the level of burial of the trap changes as a result of further subsidence or erosion, may also play a role. Examples of chemical alteration are thermal maturation, deasphalting, and microbial degradation of the reservoir oil. Examples of physical alteration of petroleum are the preferential loss of low-boiling constituents by diffusion or the addition of new constituents to the oil-in-place by migration of these constituents from a source outside the reservoir.

The thermal maturation of petroleum in the reservoir due to the geothermal gradient is a series of disproportionate reactions resulting in the production of gases, such as methane and other light hydrocarbons (Evans *et al.*, 1971). The geothermal gradient is generally on the order of 25 to 30°C/km (15°F/1000 ft or 120°C/1000 ft, i.e. 0.015°C per foot of depth or 0.012°C per foot of depth), i.e. approximately one degree for every 100 feet below the surface. The temperature gradient that exists, while not seemingly high enough for many organic reactions, may be sufficient for the task, given geologic time. Thus, with increasing depth of the reservoir, there is a tendency for crude oil to become lighter insofar as it contains increasing amounts of low-molecular weight hydrocarbons and decreasing amounts of the higher molecular weight constituents.

Deasphalting is the precipitation of asphaltene constituents from crude oils by admixture with or dissolution in the oil of large amounts of light hydrocarbon and/or gaseous hydrocarbons that vary from methane to the various heptane isomers (Mitchell and Speight, 1973; Speight *et al.*, 1984; Speight, 2007a). Such deasphalting can occur whenever considerable amounts of the lower molecular weight hydrocarbons are generated in substantial quantities because of thermal alteration of the oil or as a result of gas incursion from secondary migration. The overall effects of *gas deasphalting* are often difficult to distinguish from those of thermal maturation because both processes usually occur concomitantly and the net change in composition is that the oils become lighter (Evans *et al.*, 1971).

The microbial alteration of crude oil and alteration due to water washing — the removal of water-soluble compounds — are common methods of petroleum alteration. Both processes are frequently observed in combination because they are both due to the action of moving subsurface water. The biodegradation of crude oil is a selective utilization of certain types of hydrocarbons by microorganisms (Evans *et al.*, 1971; Bailey *et al.*, 1973a; Bailey *et al.*, 1973b; Deroo *et al.*, 1974; Connan *et al.*, 1975).

Biodegradation and water washing can be expected wherever reservoirs are close to the surface and where they are accessible to surface-derived waters. An example of such crude oil alteration is based on crude oil samples from the western Canada sedimentary basin (Deroo *et al.*, 1974). Systematic changes have been observed, from what are believed to be normal, unaltered oils in the deeper reservoirs, to heavier oils in the more shallow reservoirs, and to drastically altered oils very close to the surface — the sandstone reservoirs impregnated with heavy oils, such as the Athabasca oil sands. There are also instances of an increase in sulfur content of the crude oil along with increasing bacterial degradation. It is not yet clear whether sulfur is added to the crude oils by the action of bacteria or whether the sulfur compounds are not attacked by the bacteria and thus remain after the bacteria have preferentially eliminated the hydrocarbon moieties.

2.2.6 Relationship of Petroleum Composition and Properties

Of all the properties, specific gravity, or American Petroleum Institute (API) gravity, is the variable usually observed. The changes may simply reflect compositional differences, such as the gasoline content or asphalt content, but analysis may also show significant differences in sulfur content or even in the proportions of the various hydrocarbon types.

Elemental sulfur is a common component of sediments, and, if present in the reservoir rock, dissolves in the crude oil and reacts slowly with it to produce various sulfur compounds and/or hydrogen sulfide, which may react further with certain components of the oil. These reactions are probably much the same as those that occur in the source bed and are presumed to be largely responsible for the sulfur content of petroleum; these reactions are accompanied by a darkening of the oil and a significant rise in specific gravity and viscosity.

The variation in the character of crude oil with depth of burial is also of interest. An increase in the lighter constituents and a decrease in the density and a decrease in the amount of higher molecular weights constituents with increasing depth of the reservoir often occurs. Thus, there is the distinct possibility, if not certainty, that the original character of petroleum can be substantially altered by events following the migration process. Each of the suggested mechanisms for this alteration (i.e. thermal alteration, deasphalting, and biogenic alteration) may all play a role in the alteration of a particular crude oil. As for the choice of a migration mechanism, the choice of the prevalent in situ alteration process depends upon the prevalent conditions to which the petroleum is subject. A particular mechanism for alteration may be the prime operative, or all may contribute to determining the ultimate character of the petroleum. The determining factor is *site specificity*.

2.3 Reservoir Classification

Briefly, a reservoir is a porous and permeable underground formation containing an individual and separate natural accumulation of producible hydrocarbons (oil and/or gas), which is confined by impermeable rock or water barriers and is characterized by a single natural pressure system (Chapter 2).

A conventional petroleum reservoir (i.e. an anticline) is a dome or vault of impermeable rock, formed by the folding or faulting of the rock layers or by the rise of salt domes, with permeable rocks beneath it. Generally, there is a layer of natural gas at the top of the dome, with petroleum below, and water or brine beneath the oil. However, not all reservoirs have the above characteristics, and there are also unconventional or continuous reservoirs in which the oil is not trapped as described above. Therefore, reservoir characterization is an essential part of petroleum technology and offers an understanding or indication of the applicable method(s) of recovery.

Petroleum reservoir characterization is the process of identifying and quantifying those properties of a given petroleum reservoir that affect the distribution and migration of fluids within that reservoir. These aspects are controlled by the geological history of the reservoir. Furthermore, the ultimate goal of a hydrocarbon reservoir characterization study is the development of a reasonable physical description of a given reservoir. This physical description can then be used as a basis for simulation studies, which, in turn, are used to assess the effectiveness of various recovery strategies. An accurate physical description of the reservoir often will lead to the maximum production of hydrocarbons from the reservoir leading to decision on the economic viability of a reservoir.

The type of reservoir from which it originates and the method of recovery appear to be the most appropriate method(s) for classifying petroleum, heavy oil, and tar sand bitumen. Thus, a general classification scheme has been developed for petroleum reservoirs contained in US Department of Energy's tertiary-oil-recovery information system (TORIS). This resulted in the classification and description of 2,300 light-oil reservoirs (greater that 20°API), collectively containing 308 billion barrels of original oil-in-place (*OOIP*) (Ray *et al.*, 1991).

Reservoirs within the various classes are expected to manifest distinct types of reservoir heterogeneity as a consequence of their similar lithology and depositional histories. The creation of subclasses assists in the analysis of the impact of post deposition events on reservoir heterogeneity, which also leads to reservoir evaluation and the potential for economic viability of the reservoir.

In 1975, as part of the reaction to the energy crisis, the Securities and Exchange Commission (SEC) was interested only in the proved reserves (Chapter 4) and, by inference, interested only in those reservoirs which contain proved reserves.

On the basis of the SEC definition of reserves, reservoirs are known and proven if they contained known and proven reserves that could be recovered with reasonable certainty in future years under existing economic and operating conditions. Thus, reservoirs were classified as proven if economic producibility is supported by either actual production or conclusive formation test. Furthermore, the area of a reservoir that is considered proven includes the portion of the reservoir delineated by drilling. The immediately adjoining portions of the reservoir that were not yet drilled but could be reasonably judged as economically productive on the basis of available geological and engineering data were also included.

In addition, reservoirs that could produce oil economically through the application of improved recovery techniques, such as fluid injection, are included in the proven reservoir classification only when successful testing by a pilot project — or the operation of an installed program in the reservoir — provides support for the engineering analysis on which the project or program was based.

Reservoirs that were considered proven did not include reservoirs that contained crude oil that may become available from known reservoirs but is classified separately as indicated additional reserves. Unproven reservoirs also included reservoirs containing crude oil in which the recovery of the crude oil is subject to reasonable doubt because of uncertainty as to geology, reservoir characteristics, or economic factors. Of course, undrilled prospects are also included in the unproven reservoir category.

In summary, the proved developed reservoirs (as deduced from the SEC classification of reserves) are those reserves that produce oil from existing wells with existing equipment and operating methods.

Reservoirs from which oil can be recovered using improved recovery techniques are included only after testing by a pilot project, or after the operation of an installed program has confirmed through production response that increased recovery will be achieved.

On the other hand proved undeveloped reserves are the reservoirs from which crude oil is expected to be recovered from new wells on undrilled acreage, or from existing wells where a relatively major expenditure is required for reservoir recompletion. Undrilled reservoirs can be claimed as proven only where it can be demonstrated with certainty that there is continuity of production from the existing productive reservoir formation. Under no circumstances should estimates for proved undeveloped reservoirs be attributable to any reservoir for which an application of fluid injection or other improved recovery technique is contemplated, unless such techniques have been proved effective by actual tests in the area and in the same reservoir.

Furthermore, the Securities and Exchange Commission (SX Reg 210.4-10, November 19, 1981, as amended September 19, 1989) requires adherence to the SEC definition of proved oil and gas reserves from which classification of reservoirs can be derive.

Thus proven reservoirs (proved reservoirs) are those reservoirs where the estimated quantities of crude oil, which geological and engineering data demonstrate with reasonable certainty to be recoverable in future years from known reservoirs under existing economic and operating conditions, i.e. prices and costs as of the date the estimate is made. Prices include consideration of changes in existing prices provided only by contractual arrangements, but not on escalations based upon future conditions.

Reservoirs are considered proven if economic producibility is supported by either actual production or a conclusive formation test. The area of a reservoir considered proven includes that portion delineated by drilling and defined by gas-oil and/or oil-water contacts, if any, and the immediately adjoining portions not yet drilled, that can be reasonably judged as economically productive on the basis of available geological and engineering data. In the absence of information on fluid contacts, the lowest known structural occurrence of hydrocarbons controls the lower proved limit of the reservoir.

Reservoirs from which crude oil can be produced economically through application of improved recovery techniques, such as fluid injection, are included in the proven classification when successful testing by a pilot project, or the operation of an installed program in the reservoir, provides support for the engineering analysis on which the project or program was based.

Furthermore, reservoirs are not included in the proven category when the estimates of crude oil reserves in those reservoirs include crude oil that may become available from known reservoirs but is classified separately as indicated additional reserves, when recovery of the crude oil is subject to reasonable doubt because of uncertainty as to geology, reservoir characteristics, or economic factors, and crude oil that may occur in undrilled prospects.

Proved developed reservoirs contain proved developed oil reserves that can be recovered through existing wells with existing equipment and operating methods. These reservoirs also contain additional crude oil that is obtained through the application of fluid injection or other improved recovery techniques for supplementing the natural forces and mechanisms of primary recovery. However, such reservoirs must have been successfully tested for production of crude oil by a pilot project, or after the operation of an installed program has confirmed through production response that increased recovery will be achieved from the reservoir.

Proved undeveloped reservoirs are reservoirs that contain proved undeveloped oil reserves that are expected to be recovered from new wells on undrilled acreage, or from existing wells where a relatively major expenditure is required for recompletion. However, reservoirs that are undrilled are limited to those reservoirs that are reasonably certain of production when drilled. Proved reserves for other undrilled reservoirs can be claimed only where it can be demonstrated with certainty that there is continuity of production from the existing productive reservoir formation. Under no circumstances should a reservoir containing proved undeveloped reserves be included in the reserve estimate unless an application of fluid injection or other improved recovery technique is contemplated and unless such techniques have been proved effective by actual tests in the area and in the same reservoir.

Reservoirs that contain probable crude oil reserves are reservoirs that contain estimated quantities of crude oil, which geological and engineering data infer to be commercially recoverable, but there is uncertainty about this data precluding the classification of these reserves as proven.

The degree of risk in relying on reservoir capabilities based on estimates of probable reserves is greater than the risk for reservoirs containing estimates of proven reserves.

Finally, reservoirs that contain possible crude oil reserves are those reservoirs in which the estimated quantities of crude oil are based on limited geological and engineering data, and are commercially recoverable but there is uncertainty about this data precluding the classification of these reserves, and the reservoirs, as probable. The degree of risk in relying on reservoir capabilities based on estimates of possible reserves is greater than the risk for reservoirs containing estimates of probable reserves.

2.4 Reservoir Evaluation

One of the basic requirements of the preliminary studies of the reservoir is to provide certifications of oil and gas reserves and resources — or reservoir evaluation. Geological and geochemical analyses of hydrocarbons and organic matter in source rocks are critical in hydrocarbon exploration and in examining the extent and continuity of hydrocarbon reservoirs. Hydrocarbons may be correlated with source rock to verify the results of basin modeling.

The estimation of crude oil reserves (Chapter 1) is usually subdivided into two components: (1) the total reserves contained in the reservoir, usually termed the resource, and (2) recoverable reserves — the portion of the resource that is, as of the date the reserves are calculated, economically efficient given market conditions and rational use of modern extraction equipment and technologies. The reserve estimate will change with the price of oil and with the evolution of new and more efficient recovery technologies.

Thus, the first and foremost goal of operators is to define a reservoir with the least number of wells as possible. Drill operators drill need to obtain as much as data as possible to define the reservoir parameters and extrapolate them spatially across the field to economically evaluate the reservoir and plan the best production method.

The end point is knowledge that will lead to lower production costs and increased recoverable reserves.

2.4.1 Depletable and Renewable Resources

There are several separate concepts used to classify the stocks of depletable resources.

A depletable resource, is a resource (crude oil and natural gas in the current context) for which the natural replenishment feedback loop can be safely ignored. With depletable resources, storage, such as the strategic petroleum reserve, provides a way of extending the economic life of a resource. On the other hand, a renewable resource (e.g. biomass) is a resource with a natural replenishment rate that augments its own stock (or biomass) at a non-negligible rate.

Current reserves are those resources that can profitably be extracted at current prices. Even though current reserve is usually given as a number, it usually fluctuates with prices. Potential reserves refer to the relationship between the market price for the resource and the amount of the resource that can be profitably extracted at that price. The higher the price, the greater the potential reserves.

Resource endowment is the natural occurrence of resources in the earth's crust, which provides an upper boundary on the available (terrestrial) resources.

Finally, the static reserve index is the ratio of current reserves to current consumption and is expressed in terms of the number of years a given resource will last. However, this index is accurate if and only (1) the consumption rate continues at its current level, and (2) no new additions to reserves occur during the intervening period.

2.4.2 Development of Resources

In the extrapolation and development of petroleum resources, the properties of interest are the porosity, saturation, chemistry, and mobility. These are in pursuit of the following information:

1. The porosity of the reservoir rock and whether or not there sufficient porosity for the fluid to exist,
2. The types of fluids in the reservoir,
3. The extent of the oil saturation, i.e. the extent to which the porosity of filled with oil,
4. The mobility of the oil.

None of the above can be measured directly. They are measured in conjunction with some other parameters. For example, the injection of an electric current and the measurement of voltage response determine the electrical resistivity — the property that describes the ability if a material to support the process of charge transport. Archie's law describes the relationship between resistivity (electrical), porosity, fluid saturation, and the cementitious nature of the rock. In petroleum reservoir evaluation, this data has significant impacts and consequences for the extraction of oil and hundreds of millions of dollars may be at risk and subject to the interpretation of this data.

The current trend for reservoir evaluation is for fuller disclosure to provide investors with a better understanding of the underlying risk factors and range of uncertainty in reserve values and production forecasts.

For example, a company may have exploration potential that can represent a significant upside, but the hydrocarbons associated with the reservoirs cannot be included within the constraints of the securities exchange rules for standard reserve evaluations. Evaluators cannot create reserves. It is recommended that a separate report be published that evaluates potential reserves and provides an independent opinion as to the upside exploration potential for undrilled lands. This value reporting concept advocates providing the investor with more than the minimal information required for public disclosure. For example, a reserve may be stranded due to a lack of facilities. No economic value is given, however, there is value to the investor who has a long-term plan and can wait several years for the infrastructure to be developed. In fact, to some extent, annual reports and press releases are value reporting. An independent report would provide greater certainty to the investor and a firmer share price to the oil company.

2.4.3 New Evaluation Technology

Straight-line methods of evaluation (i.e. decline analysis) have limited applicability. In particular, decline analysis assumes pseudo steady-state flow of crude oil from the well. Many models are developed with this in mind, but the objectives of the study are not always clearly defined before any work begins. It is easy to fall into the trap of building a model that attempts to incorporate every aspect of a crude oil gathering system so that everyone can use it to forecast any possible scenario. The result is usually a model that is always waiting on more data, or one that is so complex and demanding of input data — especially revisions — that the quality of the model suffers to satisfy the quantity of input data.

Defining a set of objectives is not always an easy task. Specific circumstances can radically change the methodology required. The process of making these decisions requires a thorough understanding of the theoretical and practical aspects of reserves and deliverability.

2.5 Estimation of Reserves in Place

Oil is used by more than 200 nations of the world, but only approximately 40 countries, most of which are oil exporters, produce significant amounts. The precise numbers are likely to change because of the depletion of the once-vast resources of North America and South America deriving from the increasing domestic use of oil by many of the exporters. The number of oil exporters outside the Middle East and the former Soviet Union will decrease in the coming decades, which in turn will greatly reduce the supply diversity to the oil-importing nations (Hallock *et al.*, 2004). An increase in reliance on oil from Persian Gulf countries, West African countries (especially Nigeria), the former Soviet Union has many strategic, economic, and political implications.

Universally accepted definitions have not been developed for the many terms used by geologists, engineers, accountants, and others to denote various components of crude oil resources because most of these terms describe estimated and therefore uncertain, rather than measured, quantities. Particularly common is a general lack of understanding of the substantial difference between the terms reserves and resources (Chapter 1) as indicated by the frequent misuse of either term in place of the other.

The amount of oil in a subsurface reservoir is called oil in place (OIP) and only a fraction of this oil can be recovered from a reservoir and this fraction is called the recovery factor. The portion that can be recovered is considered to be the reserve, while the portion of the oil that is non-recoverable (residual oil; Chapter 1) should not be included unless or until methods are implemented to produce it.

There are a number of different methods of calculating oil reserves. These methods can be grouped into three general categories: (1) volumetric methods, (2) materials balance method, and (3) the decline curve method or production performance method.

Volumetric methods attempt to determine the amount of oil in place by using the size of the reservoir as well as the physical properties of its rocks and fluids. Then a recovery factor is assumed, using assumptions from fields with similar characteristics. The amount of oil in place is multiplied by the recovery factor to arrive at a reserve number. Current recovery factors for oil fields around the world typically range between 10 and 60 percent; some are over 80 percent. The wide variance is due largely to the diversity of fluid and reservoir characteristics for different deposits. The method is most useful early in the life of the reservoir, before significant production has occurred.

The materials balance method for an oil field uses an equation that relates the volume of oil, water, the gas that has been produced from a reservoir, and the change in reservoir pressure to calculate the remaining oil. It assumes that as fluids from the reservoir are produced, there will be a change in the reservoir pressure that depends on the remaining volume of oil and gas. The method requires extensive pressure-volume-temperature analysis and an accurate pressure history of the field. It requires some production to occur (typically 5% to 10% of ultimate recovery), unless reliable pressure history can be used from a field with similar rock and fluid characteristics.

The decline curve method uses production data to fit a decline curve and estimate future oil production. The three most common forms of decline curves are exponential, hyperbolic, and harmonic. It is assumed that the production will decline on a reasonably smooth curve, and so allowances must be made for shut-in wells and production restrictions. The curve can be expressed mathematically or plotted on a graph to estimate future production. It has the advantage of implicitly including all reservoir characteristics, and it requires a sufficient history to establish a statistically significant trend, ideally when production is not curtailed by regulatory or other artificial conditions.

Generally, the initial estimates of the size of newly discovered oil fields are too low. As years pass, successive estimates of the ultimate recovery of fields tend to increase. The term reserve growth refers to the typical increases in estimated ultimate recovery that occurs as oil fields are developed and produced.

Reserve growth has now become an important part of estimating total potential reserves of an individual province or country. As the world's known petroleum reserves continue to decline, there will be more pressure on geologists and engineers in the oil industry to make the reserve estimates more precise through application of the reserve-growth concept. In fact, the concept could be applied even to the undiscovered resources with some qualifications as to the inherent risk.

The system of country production quotas was introduced in the 1980s, partly based on reserves levels, and there have been dramatic increases in reported reserves among OPEC producers. In 1983, Kuwait increased its proven reserves from 67×10^9 barrels to 92×10^9 barrels. In 1985–86, the United Arab Emirates almost tripled its reserves from 33×10^9 barrels to 97×10^9 barrels. Saudi Arabia raised its reported reserve number in 1988 by 50%. From 2001 to 2002, Iran raised its proven reserves by some 30% to 130×10^9 barrels, which advanced it to second place in reserves and ahead of Iraq. Iran denied accusations of a political motive behind the readjustment, attributing the increase instead to a combination of new discoveries and improved recovery. No details were offered of how any of the upgrades were calculated, leaving doubt about the validity of these new reserve estimates (Campbell, 1977; Campbell and Laherrère, 1998).

The sudden revisions in OPEC reserves, totaling nearly 300×10^9 barrels, were surprising; this increase is partly defended by the shift in ownership of reserves away from international oil companies — some of whom were obliged to report reserves under conservative US Securities and Exchange Commission rules.

All major OPEC oil producing countries increased their reserves considerably, despite the fact that there were no new corresponding discoveries reported in this period. The reason given for the re-evaluation of reserves was that the reserve assessments in the past were too low. This may be justifiable because before the nationalization of the oil industry in these countries, private companies perhaps had a tendency to under-report reserves for financial and political reasons. In addition, OPEC production quotas are set according to reserves and also other factors which give an incentive for each country to defend their quota by keeping up with reserves. Such reserves are referred to some critics as political reserves in this context (Zittel and Schindler, 2007).

In any event, the revisions in official data had little to do with the actual discovery of new reserves. Total reserves in many OPEC countries hardly changed in the 1990s. Official reserves in Kuwait, for example, were unchanged at 96.5×10^9 barrels even though the country produced more than 8×10^9 barrels, and they did not make any important new discoveries during that period. The case of Saudi Arabia is also striking, with proven reserves estimated at between 260 and 264 × 10^9 barrels in the past 18 years — a variation of less than 2%.

2.6 Reserves

Furthermore, it is very rare that petroleum (the exception being tar sand deposits) does not occur without an accompanying cover of gas (Figure 2.2). It is therefore important when describing reserves of petroleum to also acknowledge the occurrence, properties, and character of the gaseous material, more commonly known as natural gas.

At present, oil supplies approximately 40% (natural gas is 25%) of the world's non-solar energy, and most future assessments indicate that the demand for oil will increase substantially. Predictions of impending oil shortages are as old as the industry itself, and the literature is full of arguments between optimists and pessimists about how much oil there is and what other resources may be available. There are four factors that affect the economics of oil production and its use that need to be understood to assess the availability of oil for the future:

1. The quality of the reserves
2. The quantity of the reserves
3. The likely patterns of exploitation of the resource over time
4. Those who derive benefits from the oil.

The majority of crude oil reserves identified to date are located in a relatively small number of very large fields, known as giants. In fact, approximately three hundred of the largest oil fields contain almost 75% of the available crude oil. Although most of the world's nations produce at least minor amounts of oil, the primary concentrations are in Saudi Arabia, Russia, the United States (chiefly Texas, California, Louisiana, Alaska, Oklahoma, and Kansas), Iran, China, Norway, Mexico, Venezuela, Iraq, Great Britain, the United Arab Emirates, Nigeria, and Kuwait. The largest known reserves are in the Middle East.

The defining feature of global energy markets remains high and volatile prices, reflecting a tight balance of supply and demand. This has put issues such as energy security, energy trade, and alternative energies at the forefront of the political agenda worldwide. At such a time, reliable data is an invaluable tool for decision makers and analysts both inside and outside the industry. World economic growth has faltered as a result of fluctuation in energy prices and the market turmoil that began in the summer of 2008 has continued and is likely to continue for the next year or so.

The oil price has been on an upward path for more than six years and the hiccup of the current downward price spiral highlights the interconnected nature of the energy markets. In fact, it is only a matter of time before energy-consuming countries decide that the way around this is to produce energy from other sources. And, this has already commenced. Once this happens is a substantial way, OPEC will be at the mercy of the oil consumers.

Crude oil reserves are generally taken to be those quantities that geological and engineering information indicates with reasonable certainty can be recovered in the future from known reservoirs under existing conditions, but which also include gas condensate and natural gas liquids (NGLs) as well as crude oil (BP, 2010). This data may not necessarily meet the definitions, guidelines, and practices used for determining proved reserves at company level as published by the US Securities and Exchange Commission (BP, 2010).

The natural gas reserves estimates have been compiled using a combination of primary official sources and third party data from Cedigaz and the OPEC Secretariat, and, again, the data may not necessarily meet the definitions, guidelines, and practices used for determining proved reserves at company level as published by the US Securities and Exchange Commission (BP, 2010).

2.6.1 Conventional Petroleum

Petroleum is a naturally occurring mixture of hydrocarbons, generally in a liquid state, which may also include compounds of sulfur nitrogen oxygen metals and other elements (Chapter 1) (Hsu and Robinson, 2006; Speight, 2007a, 2011).

Thus, petroleum and the equivalent term crude oil, cover a wide assortment of materials consisting of mixtures of hydrocarbons and other compounds containing variable amounts of sulfur, nitrogen, and oxygen, which may vary widely in specific gravity, API gravity, and the amount of residuum (Hsu and Robinson, 2006; Speight, 2007, 2011). Metal-containing constituents, notably those compounds that contain vanadium and nickel, usually occur in the more viscous crude oils in amounts up to several thousand parts per million, and can have serious consequences during processing of these feedstocks. Because petroleum is a mixture of widely varying constituents and proportions, its physical properties also vary widely and the color varies from near colorless to black.

At the time of writing, world crude oil reserves are on the order of 1.3 trillion barrels (1.3×10^{12} barrels) of which the United States has 28.4 billion barrels (28.4×10^9 barrels), or 2.1% of the total world reserves (BP, 2010).

In the crude state petroleum has minimal value, but when refined it provides high-value liquid fuels, solvents, lubricants, and many other products. Crude petroleum can be separated into a variety of different generic fractions by distillation. And, the terminology of these fractions has been bound by utility and often bears little relationship to composition.

The United States imports approximately 65% of its daily crude oil and crude oil product requirements. As recent events have shown, there seems to be little direction in terms of stability of supply or any measure of self-sufficiency in liquid fuels precursors, other than resorting to military action. This is particularly important for the United States refineries because a disruption in supply could cause major shortfalls in feedstock availability.

In keeping with the preferential use of lighter crude oil as well as the maturation effect in the reservoir, crude oil currently available to the refinery is somewhat different in composition and properties to those available approximately 50 years ago (Swain, 1991, 1993, 1997, 2000). The current crude oils are somewhat heavier insofar as they have higher proportions of non-volatile (asphaltic) constituents. In fact, by the standards of yesteryear, many of the crude oils in use would have been classified as heavy feedstocks, bearing in mind that they may not approach the definitions used today for heavy crude oil. Changes in feedstock character, such as this tendency to heavier materials, require adjustments to refinery operations to handle these heavier crude oils to reduce the amount of coke formed during processing and to balance the overall product slate.

2.6.2 Natural Gas

Natural gas is the gaseous mixture associated with petroleum reservoirs and is predominantly methane, but does contain other combustible hydrocarbon compounds as well as non-hydrocarbon compounds (Chapter 1) (Speight, 1993; Mokhatab *et al.*, 2006;

Speight, 2007b). In fact, associated natural gas is believed to be the most economical form of ethane (Farry, 1998). Natural gas has no distinct odor and the main use is for fuel, but it can also be used to make chemicals and liquefied petroleum gas.

At the time of writing, world natural gas reserves are on the order of 6.6 quadrillion cubic feet (6.6×10^{15} cubic feet), of which the United States has 244.7 trillion cubic feet (244.7×10^{12} cubic feet) or 3.7% of the total world reserves (BP, 2010).

It should also be remembered that the total gas resource base, like any fossil fuel or mineral resource base, is dictated by economics. Therefore, when resource data is quoted, some attention must be given to the cost of recovering those resources. Most importantly, the economics must also include a cost factor that reflects the willingness to secure total, or a specific degree of, energy independence.

Two new and possibly large sources of methane that can be expected to extend the availability of natural gas are methane hydrates (also called gas hydrates) and coal bed methane (Berecz and Balla-Achs, 1983; Sloan, 1997; Gudmundsson *et al.*, 1998; Max, 2000; Sloan, 2000). Their production technologies have only recently been developed and these sources are now becoming economically competitive.

Methane-rich gases are also produced by the anaerobic decay of non-fossil organic material and are referred to as biogas. Sources of biogas include swamps, marshes, and landfills, as well as sewage sludge and manure (Speight, 2008).

Natural gas is a vital component of the world's supply of energy and it is one of the cleanest, safest, and most useful of all energy sources. However, it needs to be understood that the word gas has a variety of different uses, and meanings. Fuel for automobiles is also called gas (being a shortened version of gasoline) but that is a totally different fuel. The gas used in a barbecue grill is actually propane (C_3H_8), which, while closely associated with and commonly found in natural gas and petroleum, is not really natural gas.

2.6.3 Heavy Oil

Thus, the generic term heavy oil is often applied inconsistently to petroleum that has an API gravity of less than 20°(Chapter 1). Other definitions classify heavy oil as having API gravity less than 22°API, or less than 25°API, and usually, but not always, sulfur content higher than 2% by weight (Ancheyta and Speight, 2007).

Furthermore, in contrast to conventional crude oils, heavy oils are darker in color and may even be black. The term heavy oil has also been arbitrarily used to describe both the heavy oils that require thermal stimulation of recovery from the reservoir and the bitumen in bituminous sand (tar sand) formations from which the heavy bituminous material is recovered by a mining operation.

The Western Hemisphere has 69% of the world's technically recoverable heavy oil and 82% of the technically recoverable natural bitumen (Meyer and Attanasi, 2003). In contrast, the Eastern Hemisphere has about 85% of the world's light oil reserves.

On the assumption that heavy oil is oil with a gravity <20°API, the U.S. heavy oil resource approaches 100 billion barrels (100×10^9 barrels) of original oil in-place (OOIP). The resource is concentrated in 248 large reservoirs, holding 80 billion barrels of OOIP, primarily located in California, Alaska, and Wyoming. Numerous other states, such as Arkansas, Louisiana, Mississippi and Texas, contain significant volumes of heavy oil. Some undeveloped heavy oil resources underlie public lands, including much of the heavy oil deposits in Alaska.

The largest heavy oil accumulation is the Venezuelan Orinoco heavy-oil belt, which contains 90% of the world's heavy oil when measured on an in-place basis. In addition to extra-heavy Orinoco oil, South America has an estimated 40 billion barrels of technically recoverable heavy oil, so that, in total, 61% of the known technically recoverable heavy oil is in South America. Of the 35 billion barrels of heavy oil estimated to be technically recoverable in North America, about 7.7 billion barrels are assigned to known producing accumulations in the lower 48 States, and 7 billion barrels are assigned to the North Slope of Alaska.

Current production of heavy oil is difficult to assess because of the loose definitions that are employed. Nevertheless, using API gravity as the guide, production of heavy oil in various countries has been calculated (Table 2.3).

2.6.4 Tar Sand Bitumen

Tar sands, also variously called oil sands or bituminous sands (Chapter 1), are loose-to-consolidated sandstone or a porous carbonate rock, impregnated with bitumen — a high boiling asphaltic material with an extremely high viscosity that is immobile under reservoir conditions and vastly different to conventional petroleum

Table 2.3 Production of heavy oil by country.

	API range	Barrels/day × 1000
Brazil	11–20	250
Canada	8–18	400*
China	12–20	180
Colombia	12–20	80
Ecuador	14–20	25
Egypt	12–20	55
India	15–17	30
Indonesia	15–20	35
Iraq	18	65
Mexico	12–20	340
Oman	15–20	35
Trinidad	15–20	55
UK	11–20	200
USA	9–19	320
Venezuela	17–20	629
Yemen	19–20	15

*Includes tar sand bitumen.

Source: http://www.heavyoilinfo.com/blog-posts/world-wide-heavy-oil-production/view

(Chapter 1) (Speight, 1990, 1997, 2008). It is therefore worth noting here the occurrence and potential supply of these materials. On an international note, the bitumen in tar sand deposits represents a potentially large supply of energy. However, many of the reserves are available only with some difficulty and that optional refinery scenarios will be necessary for conversion of these materials to liquid products because of the substantial differences in character between conventional petroleum and tar sand bitumen (Table 2.4).

Table 2.4 Comparison of the properties of Tar Sand Bitumen (Athabasca) with the properties of conventional crude oil.

Property	Bitumen (Athabasca)	Crude Oil
Specific gravity	1.01–1.03	0.85–0.90
API gravity	5.8–8.6	25–35
Viscosity, cp		
38°C/100°F	750,000	<200
100°C/212°F	11,300	
Pour point,°F	>50	ca. –20
Elemental analysis (wt. %):		
Carbon	83.0	86.0
Hydrogen	10.6	13.5
Nitrogen	0.5	0.2
Oxygen	0.9	<0.5
Sulfur	4.9	<2.0
Ash	0.8	0.0
Nickel (ppm)	250	<10.0
Vanadium (ppm)	100	<10.0
Composition wt. %)		
Asphaltenes (pentane)	17.0	<10.0
Resins	34.0	<20.0
Aromatics	34.0	>30.0
Saturates	15.0	>30.0
Carbon residue (wt. %)		
Conradson	14.0	<10.0

Because of the diversity of available information and the continuing attempts to delineate the various world tar sand deposits, it is virtually impossible to present accurate numbers that reflect the extent of the reserves in terms of the barrel unit. Indeed, investigations into the extent of many of the world's deposits are continuing at such a rate that the numbers vary from one year to the next. Accordingly, the data quoted here must be recognized as approximate with the potential of being quite different at the time of publication.

The bitumen in tar sand deposits is estimated to be at least, 1.7 trillion barrels (1.7×10^{12} barrels) in the Canadian Athabasca tar sand deposits and 1.8 trillion barrels (1.8×10^{12} barrels) in the Venezuelan Orinoco tar sand deposits, compared to 1.75 trillion barrels (1.3×10^{12} barrels) of conventional oil worldwide, most of it in Saudi Arabia and other Middle Eastern countries. Eighty-one percent of the world known recoverable bitumen is in the Athabasca tar sands of Alberta, Canada. In addition, bitumen reserves under active development are also included in the official estimate of crude oil reserves for Canada.

In spite of the high estimations of the reserves of bitumen, the two conditions of vital concern for the economic development of tar-sand deposits are the concentration of the resource, or the percent bitumen saturation, and its accessibility, usually measured by the overburden thickness. Recovery methods are based either on mining combined with some further processing or operation on the oil sands in situ. The mining methods are applicable to shallow deposits, characterized by an overburden ratio (i.e. overburden depth to thickness of tar-sand deposit). For example, indications are that for the Athabasca deposit, no more than 15% of the in-place deposit is available within current concepts of the economics and technology of open-pit mining; this 10% portion may be considered as the proven reserves of bitumen in the deposit.

2.7 References

Ancheyta, J. Speight, J.G. 2007. Heavy Oils and Residua. Hydroprocessing of Heavy Oils and Residua, Jorge Ancheyta and James G. Speight (Editors). CRC-Taylor & Francis Group, Boca Raton, Florida. 2007. Chapter 1.

Bailey, N.J.L., Jobson, A.M., and Rogers, M.A. 1973a. Chem. Geol. 11: 203.

Bailey, N.J.L., Krouse, H.H., Evans, C.R., and Rogers, M.A. 1973b. Bull. Am. Assoc. Petroleum Geologists. 57: 1276.
Bailey, N.J.L., Evans, C.R., and Milner, C.W.D. 1974. Bull. Am. Assoc. Petroleum Geologists 58: 2284.
Barker, C. 1980. Problems of Petroleum Migration W.H. Roberts and R.J. Cordell (Editors) Studies in Petroleum Geology. Bulletin No. No. 10. American Association of Petroleum Geologists, Tulsa, Oklahoma.
Barker, C., and Wang L. 1988. J. Anal. Appl. Pyrolysis 13: 9.
Berecz, E. and Balla-Achs, M. 1983. Gas Hydrates. Elsevier, Amsterdam.
BP. 2010. BP Statistical Review of World Energy 2010. BP PLC, International Headquarters, St James's Square, London, United Kingdom.
Brooks, J., and Welte, D.H. 1984. Advances in Petroleum Geochemistry. Volume I. Academic Press, New York.
Califet, Y., and Oudin, J.L. 1966. Advances in Organic Geochemistry. G.D. Hobson and G.C. Speers (Editors). Pergamon Press, New York.
Campbell, C.J. 1997. Oil & Gas Journal. 95(52): 33.
Campbell, C.J. and Laherrère, J.H. 1998. The End of Cheap Oil. Scientific American. 278: 78–83.
Connan, J., Le Tran, K., and van der Weide, B. 1975. Proc. Ninth World Petroleum Congress. 2 : 171.
Deroo, G., Tissot, B., McCrossan, R.G., and Der, F. 1974. Memoir No. 3. Canadian Society of Petroleum Geologists. pp. 148 and 184.
Evans, C.R., Rogers, M.A., and Bailey, N.J.L. 1971. Chem. Geol. 8: 147.
Farry, M. 1998. Oil & Gas Journal. 96(23): 115.
Gold, T. 1984. Scientific American. 251(5): 6.
Gold, T. 1985. Ann. Rev. Energy 10: 53.
Gold, T., and Soter, S. 1980. Scientific American 242(6): 154.
Gold. T., and Soter, S. 1982. Energy Exploration Exploitation 1(1): 89.
Gold. T., and Soter, S. 1986. Chem. Eng. News. 64(16): 1.
Gudmundsson, J.S., Andersson, V., Levik, O.I., and Parlaktuna, M. 1998. Hydrate Concept for Capturing Associated Gas. Proceedings. SPE European Petroleum Conference, The Hague, The Netherlands, 20–22 October 1998.
Hsu, C.S., and Robinson, P.R. 2006. Practical Advances in Petroleum Processing. Volume 1 and Volume 2. Springer, New York.
Kenney, J., Kutcherov, V., Bendeliani, N. and Alekseev, V. 2002. The evolution of multi-component systems at high pressures: VI. The thermodynamic stability of the hydrogen–carbon system: The genesis of hydrocarbons and the origin of petroleum. Proceedings of the National Academy of Sciences of the U.S.A. 99: 10976–10981.
Kenney, J., Shnyukov, A., Krayushkin, V., Karpov, I., Kutcherov, V. and Plotnikova, I. (2001). Dismissal of the claims of a biological connection for natural petroleum. Energia 22 (3): 26–34.
Landes, K.K. 1959. Petroleum Geology. John Wiley & Sons Inc., New York.

Max, M.D. (Editor). 2000. Natural Gas in Oceanic and Permafrost Environments. Kluwer Academic Publishers, Dordrecht, Netherlands.

Meyer, R.F., and Attanasi, E.D. 2003. Heavy Oil and Natural Bitumen - Strategic Petroleum Resources. Fact Sheet 70–03. United States Geological Survey, Washington, DC. August 2003. http://pubs.usgs.gov/fs/fs070–03/fs070–03.html.

Mitchell. D.L., and Speight, J.G. 1973. Fuel 52: 149.

Mokhatab, S., Poe, W.A., and Speight, J.G. 2006. Handbook of Natural Gas Transmission and Processing. Elsevier, Amsterdam, The Netherlands.

Oil & Gas Journal. 2008. Statistics: API Imports of Crude Oil and Products. Volume 106, No. 48.

Osborne, D. 1986. Atlantic Monthly. February. p. 39.

Schowalter, T.T. 1979. Bull. Am. Assoc. Petrol. Geol. 63: 723.

Sloan, E.D. 1997. Clathrates of Hydrates of Natural Gas. Marcel Dekker Inc., New York.

Sloan, E.D. 2000. Clathrates Hydrates: The Other Common Water Phase. Ind. Eng. Chem. Res. 39: 3123–3129.

Snowdon, L.R., and Powell. T.G. 1982. Bull. Am. Assoc. Petrol. Geol. 66: 775.

Speight, J.G. (Editor). 1990. Fuel Science and Technology Handbook, Marcel Dekker, New York.

Speight, J.G. 1993. Gas Processing: Environmental Aspects and Methods. Butterworth-Heinemann, Oxford, England.

Speight, J.G. 1997. Kirk-Othmer Encyclopedia of Chemical Technology. 4th Edition. 23: 717.

Speight. J.G. 2007a. The Chemistry and Technology of Petroleum. 4th Edition. CRC Press, Taylor & Francis Group, Boca Raton Florida.

Speight, J.G. 2007b. Natural Gas: A Basic Handbook. GPC Books, Gulf Publishing Company, Houston, Texas.

Speight, J.G. 2008. Handbook of Synthetic Fuels. McGraw-Hill, New York.

Speight, J.G. 2011. The Refinery of the Future, Gulf Professional Publishing, Elsevier, Oxford, United Kingdom.

Speight, J.G., Long, R.B., and Trowbridge, T.R. 1984. Fuel. 63: 616.

Swain, E.J. 1991. Oil & Gas Journal. 89(36): 59.

Swain, E. J. 1993. Oil & Gas Journal. 91(9): 62.

Swain, E. J. 1997. Oil & Gas Journal. 95(45): 79.

Swain, E.J. 2000. Oil & Gas Journal. March 13.

Szatmari, P. 1989. Bull. Am. Assoc. Petrol. Geol. 73(8): 989.

Tissot, B.P., and Welte, D.H. 1978. Petroleum Formation and Occurrence, Springer-Verlag, New York.

World Energy Outlook. 2008. Global Trends to 2030. International Energy Agency, Paris, France.

Zittel, W., and Schindler, J. 2007. Crude Oil: The Supply Outlook. EWG Series No. 3/2007, Energy Watch Group, Berlin, Germany. October.

3

Exploration, Recovery, and Transportation

Recent variations in the price of petroleum and of petroleum products have once again raised controversy and public displeasure. The increasing and decreasing trends in oil prices will continue and consumers must face the problem of continued oil price variations and the inevitable economic problems they cause. Price increases cause economic hardship to the consumer, while price decreases may seem to be a form of relief to the consumer, but there is another often unseen cost: the elimination of jobs.

With the estimated exhaustion of known oil reserves in about 30 to 50 years, there may be fluctuations in prices and periods of price declines, but the overall upward trend is unlikely to be reversed. There is the possibility that higher oil prices make it economical to drill for more oil at a higher cost and thereby increase supplies. However, it is not always recognized by economists and technical persons alike that finding and producing new oil may require a gap of several years, during which time field development is proceeding in a logical and non-wasteful manner.

It must be recognized that it is not appropriate to deal with the issue of petroleum pricing as an isolated entity. Petroleum pricing is a summation of many variables and must be considered in the light of total activities of the petroleum industry. This includes the processes of exploration, extraction, and transportation (as well as refining capacity, refining processes, and petroleum products such as gasoline, which are covered elsewhere; Chapter 6). Each can have an effect on crude oil prices. For example, higher prices can result in increased exploration and production both in the OPEC (oil exporting) nations and in the non-OPEC (oil importing) nations.

On the supply side, the main effects on the crude oil market are the OPEC nations, which currently provide approximately 40% of the world supply and hold approximately 70% of the proven reserves. OPEC, as the marginal supplier, behaves as a cartel by aiming to maintain excess extraction capacity in order to influence crude oil prices. In recent years, the policy has been to balance the market while allowing for an appropriate level of crude oil inventories in non-OPEC nations that,have relatively limited reserves and spare capacity.

Even before the economics of crude oil pricing is decided, there are several aspects of petroleum technology that need to be taken into account and all make a contribution to the price of oil. In fact, crude oil prices behave much as any other commodity with wide price swings but which may, or may not, be due to shortage or oversupply. Indeed, the crude oil price cycle may extend over several months or even over several years responding to changes in demand as well as supply.

Many economists are not aware of the make-up of the industry and, therefore, the means by which prices are set. On the other hand, the technical person may be aware of the processes used to discover and recover oil but may not be aware of the causes of price swings in the crude oil market and cannot appreciate the whole scenario of oil pricing.

This chapter is designed to afford a non-technical explanation of the technical aspects of oil exploration and recovery and is intended for the technical and non-technical person alike.

3.1 Exploration

Exploration provides information about future exploration prospects that can lead to revised expectations about the future value of the resource and generate new expected time paths for the resource price and quantity extracted. For example, a firm may revise its expectations about the probability of successful exploration either upward or downward in response to information obtained from its current exploration. The generation of previously unanticipated information can alter the resource price, extraction, and exploration paths so that the observed price path deviates systematically from a deterministic calculation. The observed time paths for the resource price and in situ resource value may represent a combination of many different expected price paths rather than outcomes along one fully anticipated price path.

Nevertheless, the demand for crude oil is at the highest point ever and it is getting higher. Exploration is continuing, but reservoirs are not being discovered with the same frequency that they were five decades ago (Chapter 8). Although there is a considerable amount of oil remaining in reservoirs in the United States, the oil is difficult and expensive to produce. Consequently, domestic oil production has fallen to an all-time low.

Exploration for petroleum originated in the latter part of the 19th century when geologists began to map land features to search out favorable places to drill for oil. Of particular interest to geologists were outcrops that provided evidence of alternating layers of porous and impermeable rock. The porous rock, typically a sandstone, limestone, or dolomite, provides the reservoir for the petroleum; the impermeable rock, typically clay or shale, acts as a trap and prevents migration of the petroleum from the reservoir.

By the early part of the 20th century, most of the areas where surface structural characteristics offered the promise of oil had been investigated and the era of subsurface exploration for oil began in the early 1920s (Forbes, 1958). New geological and geophysical techniques were developed for areas where the strata were not sufficiently exposed to permit surface mapping of the subsurface characteristics. In the 1960s, the development of geophysics provided methods for exploring below the surface of the earth.

The type of exploration technique employed depends upon the nature of the site. In other words, and as for many environmental operations, the recovery techniques applied to a specific site are dictated by the nature of the site and are, in fact, site specific. For example, in areas where little is known about the subsurface, preliminary reconnaissance techniques are necessary to identify potential reservoir systems that warrant further investigation.

Once an area has been selected for further investigation, more detailed methods are brought into play.

The principles used are basically magnetism (magnetometer), gravity (gravimeter), and sound waves (seismograph). These techniques are based on the physical properties of materials that can be utilized for measurements and include those that are responsive to the methods of applied geophysics. Furthermore, the methods can be subdivided into those that focus on gravitational properties, magnetic properties, seismic properties, electrical properties, electromagnetic properties, properties, and radioactive properties. These geophysical methods can be subdivided into two groups: (1) those methods without depth control and (2) those methods having depth control.

In the first group, the measurements incorporate spontaneous effects from both local and distant sources over which the observer has no control. For example, gravity measurements are affected by the variation in the radius of the earth with latitude. They are also affected by the elevation of the site relative to sea level, the thickness of the earth's crust, and the configuration and density of the underlying rocks, as well as by any abnormal mass variation that might be associated with a mineral deposit. In the last stages of assessment, the interpretation always depends upon the geological knowledge of the interpreter.

In the second group of measurements (those with depth control), seismic or electric energy is introduced into the ground and variations in transmissibility with distance are observed and interpreted in terms of geological quantities. Thus, depths to geological horizons having marked differences in transmissibility can be computed on a quantitative basis and the physical nature of these horizons deduced. The accuracy, ease of interpretation, and applicability of all methods falling into this group are not the same, and there are natural and economic conditions under which the measurements of the first group are preferable for exploration studies despite their inherent limitations.

However, it must be recognized that geophysical exploration techniques cannot be applied indiscriminately. Knowledge of the geological parameters likely to be associated with the mineral or subsurface condition being studied is essential both in choosing the method to be applied and in interpreting the results obtained. Furthermore, not all the techniques described here may be suitable for petroleum exploration. Nevertheless, the techniques that are described here are included since it is valuable to know their nature and how they might be applied to subsurface exploration.

It should also be noted that such terms as geophysical borehole logging may imply the use of one or more of the geophysical exploration techniques. This procedure involves drilling a well and using instruments to log or make measurements at various levels in the hole by such means as gravity (density), electrical resistivity, or radioactivity. In addition, formation samples (cores) are taken for physical and chemical tests.

Exploration is an economic activity and, like other economic activities, marginal benefit and marginal cost play a key role in determining how much exploration occurs. Expenditures on exploration and development reduce current profit with an expected return of adding valuable reserves for future exploitation. The efficient level of exploration activity balances the expected marginal cost of exploration with the expected benefit of exploration.

The expected marginal benefit of exploration is not just the discovery and development of new reserves but comes from the discovery of reserves whose cost of development and extraction is less than or equal to the cost of the reserves currently being depleted. In other words, the marginal value of new reserves equals the user cost of current resource depletion along an efficient time path of extraction and new reserve development.

In addition, when reserves are allowed to vary with economic and technological conditions, the upward movement of prices may no longer hold. Depending on the degree to which current extraction costs are rising as a consequence of depletion and the cost of new reserve additions, including their user cost, it is possible for new discoveries to outstrip depletion. The ability to enlarge reserves through technological breakthroughs that increase recoverability and lower the costs of extraction from existing deposits, or make new deposits easier to find and access (lowering reserve replacement costs), likewise could exert downward pressure on prices, possibly for a long period of time. However, given the uncertainties surrounding the exploration process the assumption that reserves are found on average in order of increasing cost and can be found in infinitely divisible amounts is a very serious oversimplification.

3.2 Drilling

Drilling is the final stage of the exploratory program and is in fact the only method by which a petroleum reservoir can be conclusively identified. However, in keeping with the concept of site specificity, drilling may be the only option in some areas for commencement of the exploration program. The risk involved in the drilling operation depends upon previous knowledge of the site subsurface. Thus, there is the need to relate the character of the exploratory wells at a given site to the characteristics of the reservoir.

Generally the first stage in the extraction of crude oil is to drill a well into the underground reservoir. Often many wells called multilateral wells will be drilled into the same reservoir, to ensure that the extraction rate will be economically viable. Also, some wells known as secondary wells may be used to pump water, steam, acids, or various gas mixtures into the reservoir to raise or maintain the reservoir pressure, and so maintain an economic extraction rate.

Drilling for oil is a complex operation and has evolved considerably over the past 100 years. The older cable tool method, used almost extensively until 1900, involves raising and dropping a heavy bit and drill stem attached by cable to a cantilever arm at the surface. It pulverizes the rock and earth, gradually forming the hole. The cable tool system is generally preferred only for penetrating hard rock at shallow depths and when oil reservoirs are expected at the shallow depths. The weight of the column is usually enough to attain penetration but can be augmented by a hydraulic pressure cylinder at the surface.

3.2.1 Preparing to Drill

Once the site has been selected, it must be surveyed to determine its boundaries, and environmental impact studies may need to be performed. Lease agreements, titles, and right of way accesses for the land must be obtained and evaluated legally. For offshore sites, legal jurisdiction must be determined and drilling can only commence when these issues have been settled.

Once the land has been prepared, several holes must be dug to make way for the rig and the main hole. A rectangular pit, or cellar, is dug around the location of the actual drilling hole. The cellar provides a workspace around the hole, for the workers and drilling accessories. The crew then begins drilling the main hole, often with a small drill truck rather than the main rig. The first part of the hole is larger and shallower than the main portion, and is lined with a large-diameter conductor pipe. Additional holes are dug off to the side to temporarily store equipment after which the rig equipment can be brought in and set up.

3.2.2 The Drilling Rig

Depending upon the remoteness of the drill site and its access, equipment may be transported to the site by truck, helicopter, or barge. Some rigs are built on ships or barges for work on inland water where there is no foundation to support a rig (as in marshes or lakes). Once the equipment is at the site, the rig is set up (Figure 3.1).

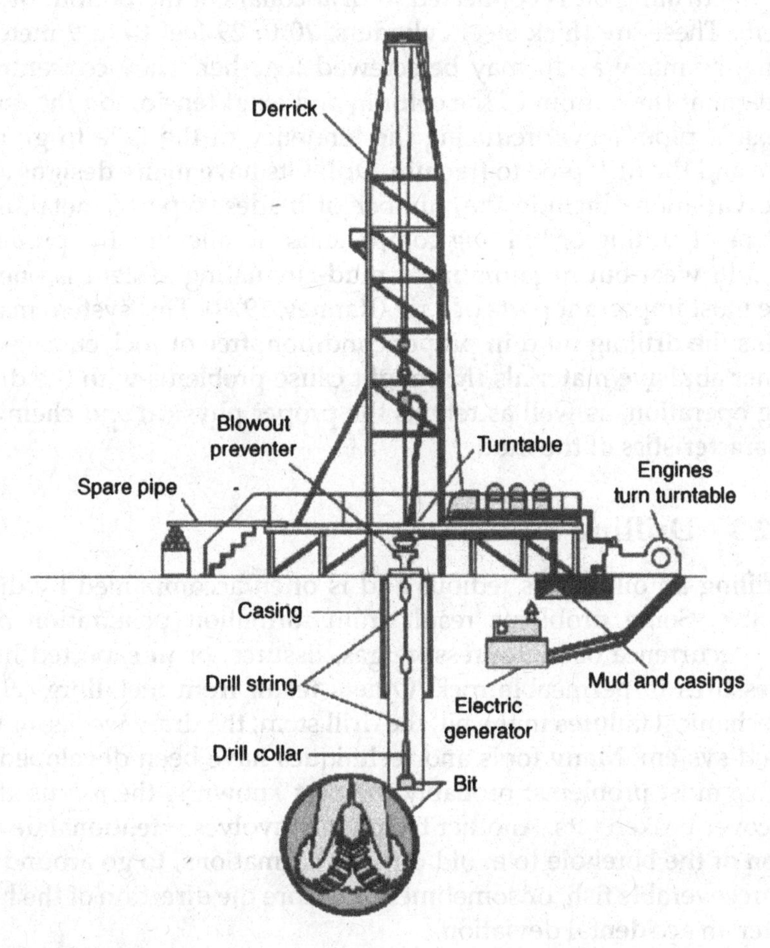

Figure 3.1 Schematic of a drilling rig.

Although there are many variations in design, all modern rotary drilling rigs have essentially the same components (Figure 3.1). The hoisting, or draw works, raises and lowers the drill pipe and casing, which can weigh as much as 200 tons (200,000 kg). The height of the derrick depends on the number of joints of drill pipe to be withdrawn as a unit before being unscrewed. The rotary table located in the middle of the rig floor rotates the drill column. The table also gaps the drill stem when the hoist is disconnected and pipe sections are inserted or removed. It imparts rotary motion to the drill stem through the kelly attached to the upper end of the column. The kelly fits into a shaped hole in the center of the rotary table. The couplings between pipe sections are called tool joints.

The drilling bit is connected to drill collars at the bottom of the stem. These are thick steel cylinders, 20 to 29 feet (6 to 9 meters) long; as many as 10 may be screwed together. They concentrate weight at the bottom of the column and exert tension on the more flexible pipe above, reducing the tendency of the hole to go off-line and the drill pipe to fracture. Drill bits have many designs and the variations include the number of blades, type of metal, and shape of cutting or brading components. To alleviate the problem of dull wear-out or jamming, a mud-circulating system is one of the most important parts of a rig (Ranney, 1979). This system maintains the drilling mud in proper condition, free of rock cuttings or other abrasive materials that might cause problems with the drilling operation, as well as retains the proper physical and chemical characteristics of the mud.

3.2.3 Drilling

Drilling an oil well is tedious and is often accompanied by difficulties. Some problems result from formation penetration and the occurrence of high-pressure gas, fissures, or unexpected high pressures in permeable rock. Others result from metallurgical or mechanical failures in the bit, the drill stem, the draw works, or the mud system. Many tools and techniques have been developed to solve most problems; probably the best known is the *fish* used to recover broken bits. Another technique involves intentional deviation of the borehole to avoid difficult formations, to go around an unrecoverable fish, or sometimes to restore the direction of the hole after an accidental deviation.

When the pre-set depth is reached, the casing pipe section are run into the hole and cemented to prevent the hole from collapsing. The casing pipe has spacers around the outside to keep it centered in the hole. The cement is pumped down the casing pipe using a bottom plug, a cement slurry, a top plug, and drilling mud. The pressure from the drilling mud causes the cement slurry to move through the casing and fill the space between the outside of the casing and the hole. Finally, the cement is allowed to harden and then tested for such properties as hardness, alignment, and a proper seal.

Drilling continues in stages and when the rock cuttings from the mud reveal the oil sand from the reservoir rock, the final depth may have been reached. At this point, the drilling apparatus is removed from the hole and tests are preformed to confirm that the final depth has been reached.

New methods to drill for oil are continually being sought, including directional or horizontal drilling techniques, to reach oil under ecologically sensitive areas, and using lasers to drill oil wells.

Directional drilling is also used to reach formations and targets not directly below the penetration point or drilling from shore to locations under water. A controlled deviation may also be used from a selected depth in an existing hole to attain economy in drilling costs. Various types of tools are used in directional drilling along with instruments to help orient their position and measure the degree and direction of deviation; two such tools are the whipstock and the knuckle joint. The whipstock is a gradually tapered wedge with a chisel-shaped base that prevents rotation after it has been forced into the bottom of an open hole. As the bit moves down, it is deflected by the taper about 5° from the alignment of the existing hole.

Drilling does not end when production commences and continues after a field enters production. Extension wells must be drilled to define the boundaries of the crude oil pool. In-field wells are necessary to increase recovery rates, and service wells are used to reopen wells that have become clogged. Additionally, wells are often drilled at the same location but to different depths, to test other geological structures for the presence of crude oil.

3.2.4 Well Completion

Once the final depth has been reached, the well is completed to allow oil to flow into the casing in a controlled manner.

First, a perforating gun is lowered into the well to the production depth. The gun has explosive charges to create holes in the casing through which oil can flow. After the casing has been perforated, a small-diameter pipe (tubing) is run into the hole as a conduit for oil and gas to flow up the well and a packer is run down the outside of the tubing. When the packer is set at the production level, it is expanded to form a seal around the outside of the tubing. Finally, a multi-valve structure (the Christmas tree; Figure 3.2) is installed at the top of the tubing and cemented to the top of the casing. The Christmas tree allows them to control the flow of oil from the well.

Tight formations are occasionally encountered and it becomes necessary to encourage flow. Several methods are used, one of which involves setting off small explosions to fracture the rock. If the formation is mainly limestone, hydrochloric acid is sent down the hole to dissolve channels in the rock. The acid is inhibited to protect the steel casing. In sandstone, the preferred method is hydraulic fracturing.

A fluid with a viscosity high enough to hold coarse sand in suspension is pumped at very high pressure into the formation,

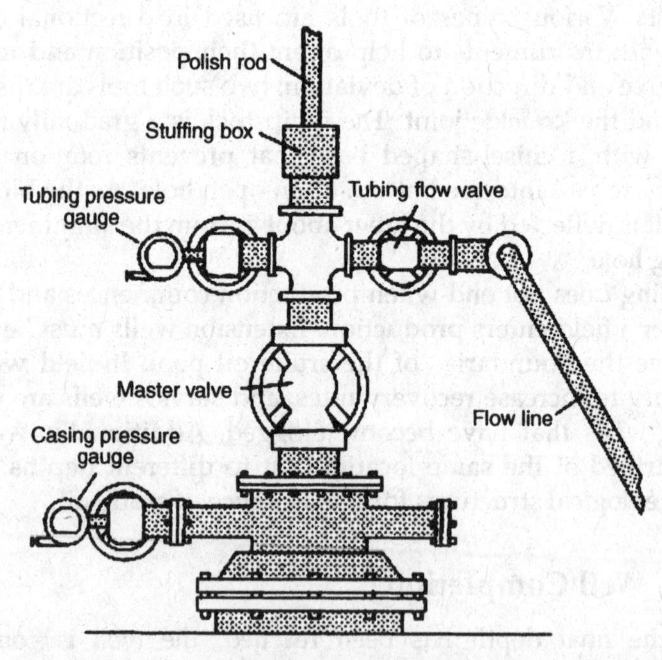

Figure 3.2 The christmas tree: A collection of control valves at the wellhead.

fracturing the rock. The grains of sand remain, helping to hold the cracks open.

Thus, once the well is completed, the flow of oil into the well is commenced. For limestone reservoir rock, acid is pumped down the well and out the perforations. The acid dissolves channels in the limestone that lead oil into the well. For sandstone reservoir rock, a specially blended fluid containing proppants (sand, walnut shells, aluminum pellets) is pumped down the well and out the perforations. The pressure from this fluid makes small fractures in the sandstone that allow oil to flow into the well, while the proppants hold these fractures open. Once the oil is flowing, the oil rig is removed from the site and production equipment is set up to extract the oil from the well.

3.3 Recovery

Recovery, as applied in the petroleum industry, is the production of oil from a reservoir. There are several methods by which this can be achieved that range from recovery due to reservoir energy (i.e. the oil flows from the well hole without assistance) to enhanced recovery methods in which considerable energy must be added to the reservoir to produce the oil. However, the effect of the method on the oil and on the reservoir must be considered before application.

Generally, crude oil reservoirs exist with an overlying gas cap, in communication with aquifers, or both. The oil resides together with water and free gas in very small holes and fractures. If the underground pressure in the oil reservoir is sufficient, the oil will be forced to the surface under this pressure (primary recovery). Natural gas (associated natural gas) is often present, which also supplies needed underground pressure (primary recovery). In this situation it is sufficient to place an arrangement of valves (the Christmas tree) on the wellhead to connect the well to a pipeline network for storage and processing. The crude oil is typically collected from individual wells by small pipelines.

In some cases this pressure later diminishes so that the oil must be pumped from the well. Natural gas or water is sometimes pumped into the well to replace the oil that is withdrawn – this is often referred to as repressurizing the oil well.

The anatomy of a reservoir is complex and is site specific, microscopically and macroscopically. Because of the various types of accumulations and the existence of wide ranges of both rock and fluid properties, reservoirs respond differently and must be treated individually. The size, shape, and degree of interconnection of the pores vary considerably from place to place in an individual reservoir. Below the oil layer, the sandstone is usually saturated with salt water. The oil is released from this formation by drilling a well and puncturing the limestone layer on either side of the limestone dome or fold. If the peak of the formation is tapped, only the gas is obtained. If the penetration is made too far from the center, only salt water is obtained.

Over the lifetime of the well the pressure will fall, and at some point there will be insufficient underground pressure to force the oil to the surface. Secondary oil recovery uses various techniques to aid in recovering oil from depleted or low-pressure reservoirs. Sometimes pumps, such as beam pumps (horse head pumps) and electrical submersible pumps are used to bring the oil to the surface. Other secondary recovery techniques increase the reservoir's pressure by water injection, natural gas re-injection, and gas lift which injects air, carbon dioxide or some other gas into the reservoir.

Enhanced oil recovery (tertiary oil recovery, EOR) relies on methods that reduce the viscosity of the oil to increase (Speight, 2009). Tertiary recovery is started when secondary oil recovery techniques are no longer enough to sustain production. For example, thermally-enhanced oil recovery methods are recovery methods in which the oil is heated the oil and make it easier to extract; usually steam is used for heating the oil.

Conventional primary and secondary recovery processes are ultimately expected to produce about one-third of the original oil discovered, although recovery (which is reservoir and oil dependent) from individual reservoirs can range from less than 5% to as high as 80% of the original oil in place. This broad range of recovery efficiency is a result of variations in the properties of the specific rock and fluids involved from reservoir to reservoir as well as the kind and level of energy that drives the oil to producing wells, where it is captured.

Conventional oil production methods may be unsuccessful because the management of the reservoir was poor or because reservoir heterogeneity has prevented the recovery of crude oil in an economical manner. Reservoir heterogeneity, such as fractures and faults, can cause reservoirs to drain inefficiently by conventional methods. Also, highly cemented or shale zones can produce barriers to the flow of fluids in reservoirs and lead to high residual oil saturation. Reservoirs containing crude oils with low API gravity often cannot be produced efficiently without application of enhanced oil recovery (EOR) methods because of the high viscosity of the crude oil. In some cases, the reservoir pressure was depleted prematurely by poor reservoir management practices to create reservoirs with low energy and high oil saturation.

Recovery of oil when a well is first opened is usually by natural flow forced by the pressure of the gas or fluids that are contained within the deposit. There are several means that serve to drive the petroleum fluids from the formation, through the well, and to the surface, and these methods are classified as either natural or applied flow.

3.3.1 Primary Recovery (Natural Methods)

If the underground pressure in the oil reservoir is sufficient, then the oil will be forced to the surface under this pressure. Gaseous fuels or natural gas are usually present, which also supplies needed underground pressure. In this situation it is sufficient to place a complex arrangement of valves (the Christmas tree) at the wellhead to connect the well to a pipeline network for storage and processing. This is called primary oil recovery.

Thus, primary oil production (primary oil recovery) is the first method of producing oil from a well and depends upon natural reservoir energy to drive the oil through the complex pore network to producing wells. If the pressure on the fluid in the reservoir is great enough, the oil flows into the well and up to the surface. Such driving energy may be derived from liquid expansion and evolution of dissolved gases from the oil as reservoir pressure is lowered during production, expansion of free gas, or a gas cap, influx of natural water, gravity, or combinations of these effects.

Crude oil moves out of the reservoir into the well by one or more of three processes. These processes are: dissolved gas drive, gas cap drive, and water drive. Early recognition of the type of drive involved is essential to the efficient development of an oil field.

In dissolved gas drive, the propulsive force is the gas in solution in the oil, which tends to come out of solution because of the pressure release at the point of penetration of a well. Dissolved gas drive is the least efficient type of natural drive as it is difficult to control the gas-oil ratio; the bottom-hole pressure drops rapidly, and the total eventual recovery of petroleum from the reservoir may be less than 20%.

If gas overlies the oil beneath the top of the trap, it is compressed and can be utilized to drive the oil into wells situated at the bottom of the oil-bearing zone. By producing oil only from below the gas cap, it is possible to maintain a high gas-oil ratio in the reservoir until almost the very end of the life of the pool. However, if the oil deposit is not systematically developed so that bypassing of the gas occurs, an undue proportion of oil is left behind. The usually recovery of petroleum from a reservoir in a gas cap field is 40 to 50%.

Usually the gas in a gas cap contains methane and other hydrocarbons that may be separated out by compressing the gas. A well-known example is natural gasoline that was formerly referred to as casinghead gasoline or natural gas gasoline. However, at high pressures, such as those existing in the deeper fields, the density of the gas increases and the density of the oil decreases until they form a single phase in the reservoir. These are the so-called retrograde condensate pools because a decrease (instead of an increase) in pressure brings about condensation of the liquid hydrocarbons. When this reservoir fluid is brought to the surface and the condensate is removed, a large volume of residual gas remains. The modern practice is to cycle this gas by compressing it and inject it back into the reservoir, thus maintaining adequate pressure within the gas cap, and condensation in the reservoir is prevented. Such condensation prevents recovery of the oil, for the low percentage of liquid saturation in the reservoir precludes effective flow.

The most efficient propulsive force in driving oil into a well is natural water drive, in which the pressure of the water forces the lighter recoverable oil out of the reservoir into the producing wells. In anticlinal accumulations, the structurally lowest wells around the flanks of the dome are the first to come into water. Then the oil-water contact plane moves upward until only the wells at the top of the anticline are still producing oil; eventually these also must be abandoned as the water displaces the oil.

In a water drive field, it is essential that the removal rate be adjusted so that the water moves up evenly as space is made available for it by the removal of the hydrocarbons. An appreciable decline in bottom-hole pressure is necessary to provide the pressure gradient required to cause water influx. The pressure differential needed depends on the reservoir permeability; the greater the permeability, the less the difference in pressure necessary.

The recovery of petroleum from the reservoir in properly operated water drive pools may run as high as 80%. The force behind the water drive may be hydrostatic pressure, the expansion of the reservoir water, or a combination of both. Water drive is also used in certain submarine fields.

Gravity drive is an important factor when oil columns of several thousands of feet exist, as they do in some North American fields. Furthermore, the last bit of recoverable oil is produced in many pools by gravity drainage of the reservoir. Another source of energy during the early stages of withdrawal from a reservoir containing under-saturated oil is the expansion of that oil as the pressure reduction brings the oil to the bubble point (the pressure and temperature at which the gas starts to come out of solution).

For primary recovery operations, no pumping equipment is required. If the reservoir energy is not sufficient to force the oil to the surface, then the well must be pumped. In either case, nothing is added to the reservoir to increase or maintain the reservoir energy or to sweep the oil toward the well. The rate of production from a flowing well tends to decline as the natural reservoir energy is expended. When a flowing well is no longer producing at an efficient rate, a pump is installed.

The recovery efficiency for primary production is generally low when liquid expansion and solution gas evolution are the driving mechanisms. Much higher recoveries are associated with reservoirs with water and gas cap drives and with reservoirs in which gravity effectively promotes drainage of the oil from the rock pores. The overall recovery efficiency is related to how the reservoir is delineated by production wells. Thus, for maximum recovery by primary recovery, it is often preferable to sink several wells into a reservoir, thereby bringing about recovery by a combination of the methods outlined here.

3.3.2 Secondary Recovery

Over the lifetime of the well, the pressure will fall, and at some point there will be insufficient underground pressure to force the oil to the surface. If economical, and it often is, the remaining oil in the well is extracted using secondary oil recovery methods. It is at this point that secondary recovery methods must be applied.

Secondary oil recovery methods use various techniques to aid in recovering oil from depleted or low-pressure reservoirs. Sometimes pumps on the surface or submerged (electrical submersible pumps, ESPs), are used to bring the oil to the surface. Other secondary recovery techniques increase the reservoir's pressure by water injection and gas injection, which injects air or some other gas into the reservoir are used.

Together, primary recovery and secondary recovery allow 25% to 35% of the reservoir's oil to be recovered.

Primary (or conventional) recovery can leave as much as 70% of the petroleum in the reservoir. Thus, there are two main objectives in secondary crude oil production: (1) to supplement the depleted reservoir energy pressure and (2) to sweep the crude oil from the injection well toward and into the production well. In fact, secondary oil recovery involves the introduction of energy into a reservoir to produce more oil. For example, the addition of materials to reduce the interfacial tension of the oil results in a higher recovery of oil.

The most common follow-up, or secondary recovery, operations usually involve the application of pumping operations or of injection of materials into a well to encourage movement and recovery of the remaining petroleum. The pump, generally known as the horsehead pump or the sucker rod pump (Figure 3.3), provides mechanical lift to the fluids in the reservoir. The most commonly recognized oil-well pump is the reciprocating or plunger pumping equipment (also called a sucker-rod pump), which is easily recognized by the horsehead beam pumping jacks. A pump barrel is lowered into the well on a string of 6 inch (inner diameter) steel rods known as sucker rods. The up-and-down movement of the sucker rods forces the oil up the tubing to the surface. A walking beam powered by a nearby engine may supply this vertical

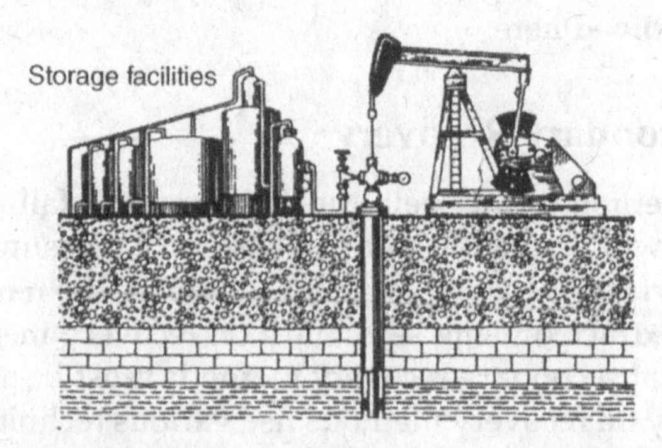

Figure 3.3 The horsehead pump.

movement, or it may be brought about through the use of a pump jack, which is connected to a central power source by means of pull rods. Electrically powered centrifugal pumps and submersible pumps — both pump and motor are in the well at the bottom of the tubing — have proven their production capabilities in numerous applications.

There are also secondary oil recovery operations that involve the injection of water or gas into the reservoir. When water is used the process is called a waterflood; with gas, the process is called a gasflood. Separate wells are usually used for injection and production. The injected fluids maintain reservoir pressure or re-pressure the reservoir after primary depletion and displace a portion of the remaining crude oil to production wells. In fact, the first method recommended for improving the recovery of oil was probably the re-injection of natural gas, and there are indications that gas injection was utilized for this purpose before 1900 (Craft and Hawkins, 1959; Frick, 1962). These early practices were implemented to increase the immediate productivity and are therefore classified as pressure maintenance projects. Recent gas injection techniques have been devised to increase the ultimate recovery, thus qualifying as secondary recovery projects.

In secondary recovery, the injected fluid must dislodge the oil and propel it toward the production wells. Reservoir energy must also be increased to displace the oil. Using techniques such as gas and water injection, there is no change in the state of oil. Similarly, there is no change in the state of the oil during miscible fluid displacement technologies. The analogy that might be used is that of a swimmer in which there is no change to the natural state of the human body.

Thus, the success of secondary recovery processes depends on the mechanism by which the injected fluid displaces the oil (displacement efficiency) and on the volume of the reservoir that the injected fluid enters (conformance or sweep efficiency). In most proposed secondary projects, water does both these things more effectively than gas. It must be decided if the use of gas offers any economic advantages because of availability and relative ease of injection. In reservoirs with high permeability and high vertical span, the injection of gas may result in high recovery factors as a result of gravity segregation, as described in a later section. However, if the reservoir lacks either adequate vertical permeability or the possibility for gravity segregation, a frontal drive similar to that used for water injection can be used (dispersed gas injection). Thus, dispersed gas injection is anticipated to be more effective in reservoirs that are relatively thin and have little dip. Injection into the top of the formation or into the gas cap is more successful in reservoirs with higher vertical permeability (200 md or more) and enough vertical relief to allow the gas cap to displace the oil downward.

During the withdrawal of fluids from a well, it is usual practice to maintain pressures in the reservoir at or near the original levels by pumping either gas or water into the reservoir as the hydrocarbons are withdrawn. This practice has the advantage of retarding the decline in the production of individual wells and considerably increasing the ultimate yield. It also may bring about the conservation of gas that otherwise would be wasted, and the disposal of brines that otherwise might pollute surface and near-surface potable waters.

Water injection is still predominantly a secondary recovery process — waterflood. Probably the principal reason for this is that reservoir formation water is ordinarily not available in volume during the early years of an oil field and pressure maintenance water from outside the field may be too expensive. When a young field produces considerable water, it may be injected back into the reservoir primarily for the purpose of nuisance abatement, but reservoir pressure maintenance is a valuable by-product.

Nevertheless, some passages in the formation are larger than others, and the water tends to flow freely through these, bypassing smaller passages where the oil remains. A partial solution to this problem is possible by miscible fluid flooding. Liquid butane and propane are pumped into the ground under considerable pressure, dissolving the oil and carrying it out of the smaller passages; additional pressure is obtained by using natural gas.

3.3.3 Enhanced Oil Recovery

Enhanced oil recovery (tertiary oil recovery) is the incremental ultimate oil that can be recovered from a petroleum reservoir over oil that can be obtained by primary and secondary recovery methods (Speight, 2009). In fact, certain reservoir types, such as those with very viscous crude oils and some low-permeability carbonate (limestone, dolomite, or chert) reservoirs, respond poorly to conventional secondary recovery techniques.

In these reservoirs it is desirable to initiate enhanced oil recovery (EOR) operations as early as possible. This may mean considerably abbreviating conventional secondary recovery operations or bypassing them altogether. Thermal floods using steam and controlled in situ combustion methods are also used. Thermal methods of recovery reduce the viscosity of the crude oil by heat so that it flows more easily into the production well. Thus tertiary techniques are usually variations of secondary methods with a goal of improving the sweeping action of the invading fluid.

Enhanced oil recovery methods are designed to reduce the viscosity of the crude oil (i.e. to reduce the pour point of the crude oil relative to the temperature of the reservoir), thereby increasing oil production. Enhanced oil recovery methods are applied started when secondary oil recovery techniques are no longer enough to sustain production. The oil remaining after conventional recovery operations is retained in the pore space of reservoir rock at a lower concentration than originally existed. In portions of the reservoir that have been contacted or swept by the injection fluid, the residual oil remains as droplets (or ganglia) trapped in either individual pores or clusters of pores. It may also remain as films partly coating the pore walls. Entrapment of this residual oil is predominantly due to capillary and surface forces and to geometry of the pore systems.

Enhanced oil recovery processes use thermal, chemical, or fluid phase behavior effects to reduce or eliminate the capillary forces that trap oil within pores, to thin the oil or otherwise improve its mobility or to alter the mobility of the displacing fluids (Speight, 2009). In some cases, the effects of gravity forces, which ordinarily cause vertical segregation of fluids of different densities, can be minimized or even used to advantage. The various processes differ considerably in complexity, the physical mechanisms responsible for oil recovery, and the amount of experience that has been derived from field application. The degree to which the enhanced oil recovery methods are applicable in the future will depend on development of improved process technology. It will also depend on improved understanding of fluid chemistry, phase behavior, and physical properties; and on the accuracy of geology and reservoir engineering in characterizing the physical nature of individual reservoirs (Borchardt and Yen, 1989).

Thermal methods for oil recovery have found most use when the oil in the reservoir has a high viscosity. For example, heavy oil is usually highly viscous with a viscosity ranging from approximately

100 centipoises to several million centipoises at the reservoir conditions (Figure 3.4). In addition, oil viscosity is also a function of temperature and API gravity (Speight, 2000). These processes add heat to the reservoir to reduce oil viscosity and/or to vaporize the oil. In both instances, the oil is made more mobile so that it can be more effectively driven to producing wells. In addition to adding heat, these processes provide a driving force to move oil to producing wells.

Thermal recovery methods include cyclic steam injection; steam flooding, and in situ combustion. The steam processes are the most advanced of all enhanced oil recovery methods in terms of field experience and thus have the least uncertainty in estimating performance, provided that a good reservoir description is available. Steam processes are most often applied in reservoirs containing viscous oils and tars, usually in place of rather than following secondary or primary methods. Commercial application of steam

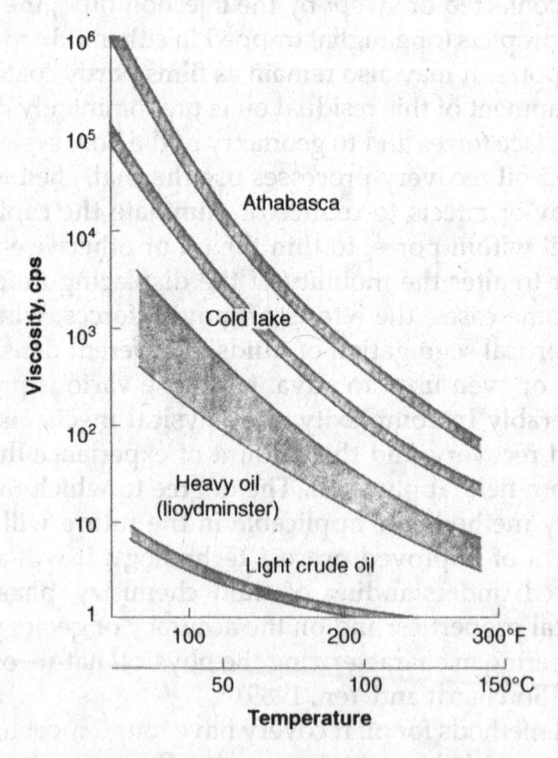

Figure 3.4 Variation of petroleum, heavy oil, and tar sand bitumen viscosity with temperature.

processes has been underway since the early 1960s. In situ combustion has been field tested under a wide variety of reservoir conditions, but few projects have proven economical and advanced to commercial scale.

Cyclic steam injection (steam soak, huff 'n' puff) is the alternating injection of steam and production of oil with condensed steam from the same well or wells. A cyclic steam injection process includes three stages. In the first stage is injection, during which a measured amount of steam is introduced into the reservoir. In the second stage (the soak period) requires that the well be shut in for a period of time of usually several days to allow uniform heat distribution to reduce the viscosity of the oil, alternatively, to raise the reservoir temperature above the pour point of the oil. Finally, during the third stage, the now-mobile oil is produced through the same well. The cycle is repeated until the flow of oil diminishes to a point of no returns.

In situ combustion is normally applied to reservoirs containing low-gravity oil but has been tested over perhaps the widest spectrum of conditions of any enhanced oil recovery process. In the process, heat is generated within the reservoir by injecting air and burning part of the crude oil. This reduces the oil viscosity and partially vaporizes the oil in place, and the oil is driven out of the reservoir by a combination of steam, hot water, and gas drive. Forward combustion involves movement of the hot front in the same direction as the injected air. Reverse combustion involves movement of the hot front opposite to the direction of the injected air.

The relatively small portion of the oil that remains after these displacement mechanisms have acted becomes the fuel for the in situ combustion process. Production is obtained from wells offsetting the injection locations. In some applications, the efficiency of the total in situ combustion operation can be improved by alternating water and air injection. The injected water tends to improve the utilization of heat by transferring heat from the rock behind the combustion zone to the rock immediately ahead of the combustion zone.

In modified in situ extraction processes, combinations of in situ and mining techniques are used to access the reservoir. A portion of the reservoir rock must be removed to enable application of the in situ extraction technology. The most common method is to enter the reservoir through a large-diameter vertical shaft, excavate horizontal drifts from the bottom of the shaft, and drill injection and production wells horizontally from the drifts. Thermal extraction processes are then applied through the wells. When the horizontal wells are drilled at or near the base of the tar sand reservoir, the injected heat rises from the injection wells through the reservoir, and drainage of produced fluids to the production wells is assisted by gravity.

3.4 Bitumen Recovery

In the case of tar sand bitumen, the alternative to in situ processing is to mine the tar sands, transport them to a processing plant, extract the bitumen value, and dispose of the waste sand. Such a procedure is often referred to as oil mining. This is the term applied to the surface or subsurface excavation of petroleum-bearing formations for subsequent removal of the heavy oil or bitumen by washing, flotation, or retorting treatments. Oil mining also includes recovery of heavy oil by drainage from reservoir beds to mine shafts or other openings driven into the rock, or by drainage from the reservoir rock into mine openings driven outside the tar sand but connected with it by bore holes or mine wells.

The API gravity of tar sand bitumen is usually less than 10° depending upon the deposit, and whereas conventional crude oils may have a viscosity of several poise (at 40°C, 105°F), the tar sand bitumen has a viscosity of the order of 50,000 centipoises to 1,000,000 centipoises or more at formation temperatures (approximately 0°C to 10°C, 32°F to 50°F depending upon the season). This offers a formidable but not insurmountable obstacle to bitumen recovery.

3.4.1 Mining Methods

The mining method of recovery has received considerable attention since it was chosen as the technique of preference for the only two commercial bitumen recovery plants in operation in North America. In situ processes have been tested many times in the United States, Canada, and other parts of the world and are ready for commercialization. There are also conceptual schemes that are a combination of both mining (above-ground recovery) and in situ (non-mining recovery) methods.

Surface mining is the mining method that is currently being used by Suncor Energy and Syncrude Canada Limited to recover oil sand from the ground. Surface mining can be used in mineable oil sand areas which lie under 75 meters (250 feet) or less of overburden material. Only 7% of the Athabasca Oil Sands deposit can be mined using the surface mining technique, as the other 93% of the deposit has more than 75 meters of overburden. This other 93% will have to be mined using different mining techniques.

The first step in surface mining is the removal of muskeg and overburden. Muskeg is a water-soaked area of decaying plant material that is one to three meters thick and lies on top of the overburden material. Before the muskeg can be removed it must be drained of its water content. The process can take up to three years to complete. Once the muskeg has been drained and removed, the overburden must also be removed. Overburden is a layer of clay, sand, and silt that lies directly above the oil sands deposit. Overburden is used to build dams and dykes around the mine and will eventually be used for land reclamation projects. When all of the overburden is removed, the oil sand is exposed and is ready to be mined.

There are two methods of mining currently in use in the Athabasca Oil Sands. Suncor Energy uses the truck-and-shovel method of mining whereas Syncrude uses the truck and shovel method of mining, as well as draglines and bucket-wheel reclaimers. These enormous draglines and bucket-wheels are being phased out and soon will be completely replaced with large trucks and shovels. The shovel scoops up the oil sand and dumps it into a heavy hauler truck. The heavy hauler truck takes the oil sand to a conveyor belt that transports the oil sand from the mine to the extraction plant.

Tar sand, as mined commercially in Canada, contains an average of 10 to 12% bitumen, 83 to 85% mineral matter and 4 to 6% water. A film of water coats most of the mineral matter, and this property permits extraction by the hot-water process, which is, to date, the only successful commercial process to be applied to bitumen recovery from mined tar sands in North America (Speight, 2008). Many process options have been tested with varying degrees of success and one of these options may even supersede the hot water process.

In the hot water extraction process, the tar sand feed is introduced into a conditioning drum. In this step the oil sand is heated, mixed water, and agglomeration of the oil particles begins. The conditioning is carried out in a slowly rotating drum that contains a steam-sparging system for temperature control as well as mixing devices to assist in lump size reduction and a size ejector at the outlet end.

One of the economic issues that arises from the hot water process is the disposal and control of the tailings. The fact is that each ton of oil sand in place has a volume of about 16 cubic feet, which will generate about 22 cubic feet of tailings giving a volume gain on the order of 40%. If the mine produces about 200,000 tons of oil sand per day, the volume expansion represents a considerable solids disposal problem.

3.4.2 Non-Mining Methods

In principle, the non-mining recovery of bitumen from tar sand deposits is an enhanced recovery technique and requires the injection of a fluid into the formation through an injection wall. This leads to the in situ displacement of the bitumen from the recovery and bitumen production at the surface through an egress (production well). However, there are several serious constraints that are particularly important and relate to bulk properties of the tar sand and the bitumen. In fact, both must be considered in toto in the context of bitumen recovery by non-mining techniques. For example, such processes need a relatively thick layer of overburden to contain the driver substance within the formation between injection and production wells.

One of the major deficiencies in applying mining techniques to bitumen recovery from tar sand deposits is (next to the immediate capital costs) the associated environmental problems. Moreover, in most of the known deposits, the vast majority of the bitumen lies in formations in which the overburden/pay zone ratio is too high. Therefore, it is not surprising that over the last two decades a considerable number of pilot plants have been applied to the recovery of bitumen by non-mining techniques from tar sand deposits where the local terrain and character of the tar sand may not always favor a mining option.

The steam processes are the most advanced of all enhanced oil recovery methods in terms of field experience and thus have the least uncertainty in estimating performance, provided that a good reservoir description is available. Steam processes are most often applied in reservoirs containing viscous oils and tars, usually in place of rather than following secondary or primary methods. Commercial application of steam processes has been underway since the early 1960s. In situ combustion has been field tested under a wide variety of reservoir conditions, but few projects have proven economical and advanced to commercial scale.

Other variations on this theme include the use of steam and the means of reducing interfacial tension by the use of various solvents. The solvent extraction approach has had some success when applied to bitumen recovery from mined tar sand but when applied to unmined material, losses of solvent and bitumen are always a major obstacle. This approach should not be rejected out of hand because a novel concept may arise that guarantees minimal but acceptable losses of bitumen and solvent.

In situ combustion is normally applied to reservoirs containing low-gravity oil but has been tested over perhaps the widest spectrum of conditions of any enhanced oil recovery process. In the process, heat is generated within the reservoir by injecting air and burning part of the crude oil. This reduces the oil viscosity and partially vaporizes the oil in place, and the oil is driven out of the reservoir by a combination of steam, hot water, and gas drive. Forward combustion involves movement of the hot front in the same direction as the injected air; reverse combustion involves movement of the hot front opposite to the direction of the injected air.

Finally, by all definitions, the quality of the bitumen from tar sand deposits is poor as a refinery feedstock. As in any field in which primary recovery operations are followed by secondary or enhanced recovery operations and there is a change in product quality, such is also the case for tar sand recovery operations. Thus, product oils recovered by the thermal stimulation of tar sand deposits show some improvement in properties over those of the bitumen in-place.

Although this improvement in properties may not appear to be too drastic, nevertheless it usually is sufficient to have major advantages for refinery operators. Any incremental increase in the units of hydrogen/carbon ratio can save amounts of costly hydrogen during upgrading. The same principles are also operative for reductions in the nitrogen, sulfur, and oxygen contents.

Many innovative concepts in heavy oil production have been developed and several major new technologies have positively affected the heavy oil industry in the last 10 years.

Steam assisted gravity drainage (SAGD) was developed first in Canada for reservoirs where the immobile bitumen occurs and uses paired horizontal wells. Low-pressure steam continuously injected through the upper well, creates a steam chamber along the walls of which the heated bitumen flows and is produced in the lower well.

Several variations of this process have been developed. One variation uses a single horizontal well, with steam injection through a central pipe and production along the annulus. Another variation involves steam injection through existing vertical wells and production through an underlying horizontal well. The key benefits of the SAGD process are an improved steam-oil ratio and high ultimate recovery (on the order of 60 to 70%). The outstanding technical issues relate to low initial oil rate, artificial lifting of bitumen to the surface, horizontal well operation and the extrapolation of the process to reservoirs having low permeability, low pressure or bottom water.

Heat generation is a major economic constraint on all thermal processes. Currently, steam is generated with natural gas, and when the cost of natural gas rises, operating costs rise considerably. Thermally, SAGD is about twice as efficient as cyclic steam stimulation, with steam-oil ratios that are now approaching two (instead of four for cyclic steam soak), for similar cases. Combined with the high recovery ratios possible, SAGD will likely displace pressure-driven thermal process in all cases where the reservoir is reasonably thick.

Vapor-assisted petroleum extraction (VAPEX) is a new process in which the physics of the process are essentially the same as for SAGD and the configuration of wells is generally similar. The process involves the injection of vaporized solvents such as ethane or propane to create a vapor-chamber through which the oil flows due to gravity drainage (Butler and Mokrys, 1991; United States Patent 5,407,009; United States Patent 5,607,016. 1995; Butler and Jiang, 2000). The process can be applied in paired horizontal wells, single horizontal wells or a combination of vertical and horizontal wells. The key benefits are significantly lower energy costs, potential for in situ upgrading and application to thin reservoirs, with bottom water or reactive mineralogy.

VAPEX can be used in conjunction with SAGD methods. A key factor is the generation of a three-phase system with a continuous gas phase so that as much of the oil as possible can be contacted by the gaseous phases, generating the thin oil film drainage mechanism. Vertical permeability barriers are a problem, and must be overcome through hydraulic fracturing to create vertical permeable channels, or undercut by lateral growth of the chamber beyond the lateral extent of the limited barrier, or baffle.

Microbial enhanced oil recovery (MEOR) processes involve the use of reservoir microorganisms or specially selected natural bacterial to produce specific metabolic events that lead to enhanced oil recovery (Speight, 2009).

Microbial enhanced oil recovery differs from chemical enhanced oil recovery in the method by which the enhancing products are introduced into the reservoir. However, even though microbes produce the necessary chemical reactions in situ whereas surface injected chemicals may tend to follow areas of higher permeability, resulting in decreased sweep efficiency, there is need for caution and astute observation of the effects of the microorganisms on the reservoir chemistry. The mechanism by which microbial enhanced oil recovery processes work can be quite complex and may involve multiple biochemical processes.

3.5 Transportation

Most oil fields are a considerable distance from the refineries that convert crude oil into usable products, and therefore the oil must be transported in pipelines and tankers (Figure 3.5). However, most crude oil needs some form of treatment near the reservoir

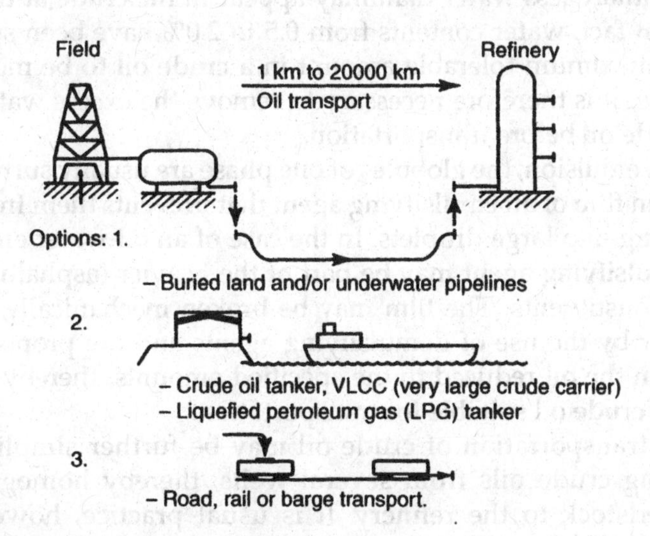

Figure 3.5 Petroleum transportation.

before it can be carried considerable distances through the pipelines or in the tankers. Railroad cars and motor vehicles are also used to a large extent for the transportation of petroleum products.

Fluids produced from a well are seldom pure crude oil. In fact, the oil often contains quantities of gas, saltwater, or even sand. Separation must be achieved before transportation. Separation and cleaning usually take place at a central facility that collects the oil produced from several wells. Gas can be separated conveniently at the wellhead. When the pressure of the gas in the crude oil as it comes out at the surface is not too great, a simple flow tank fitted with baffles can be used to separate the gas from the oil at atmospheric pressure. If a considerable amount of gas is present, particularly if the crude oil is under considerable pressure, a series of flow tanks is necessary. The natural gas itself may contain as impurities one or more non-hydrocarbon substances. The most abundant of these impurities is hydrogen sulfide, which imparts a noticeable odor to the gas. A small amount of this compound, or an equally odiferous derivative, is considered advantageous as it gives an indication of leaks and where they occur, (Mokhatab *et al.*, 2006).

Another step that needs to be taken in the preparation of crude oil for transportation is the removal of excessive quantities of water. Crude oil at the wellhead usually contains emulsified water in proportions that may reach amounts approaching 80 to 90%. It is generally required that crude oil to be transported by pipeline contain substantially less water than may appear in the crude at the wellhead. In fact, water contents from 0.5 to 2.0% have been specified as the maximum tolerable amount in a crude oil to be moved by pipeline. It is therefore necessary to remove the excess water from the crude oil before transportation.

In an emulsion, the globules of one phase are usually surrounded by a thin film of an emulsifying agent that prevents them from congregating into large droplets. In the case of an oil-water emulsion, the emulsifying agent may be part of the heavier (asphaltic) more polar constituents. The film may be broken mechanically, electrically, or by the use of demulsifying agents and the proportion of water in the oil reduced to the specified amounts, thereby tendering the crude oil suitable for transportation.

The transportation of crude oil may be further simplified by blending crude oils from several wells, thereby homogenizing the feedstock to the refinery. It is usual practice, however, to blend crude oils of similar characteristics although fluctuations in the properties of the individual crude oils may cause significant variations in the properties of the blend over a period of time. However, the technique of blending several crude oils before transportation, or even after transportation but before refining, may eliminate the frequent need to change the processing conditions that would perhaps be required to process each of the crude oils individually.

The arrival of large quantities of petroleum oil at import and refining centers has brought about the need for storage facilities. The usual form of crude oil storage is the collection of large cylindrical steel storage tanks (*tank farm*) that are a familiar sight at most refineries and shipping terminals. The tanks vary in size, but some are capable of holding up to 950,000 barrels of oil. Crude oil may also be stored in such geological features as salt domes. The domes have been previously leached or hollowed out into huge underground caves, such as those used by the US Strategic Petroleum Reserve in Louisiana and Texas. Other underground storage facilities include disused coal mines and artificial caverns. Natural gas is, on occasion, stored in old reservoirs from which the gas has been recovered. The gas is pumped under pressure into the reservoir at times of low gas demand so that it can be retrieved later to meet peak demand.

Pipelines usually accomplish large-scale transportation of crude oil, refined petroleum products, natural gas, and tankers, while smaller-scale distribution, especially of petroleum products, is carried out by barges, trucks, and rail tank cars. In fact, the transportation from the source of the crude oil to the market is as old as the industry itself. Even the bitumen used in Babylon and other cities of the Fertile Crescent had to be transported from the seepage at Hit to the place where it would be used.

In more modern times, the transportation of crude oil from fields to refineries and of products to market centers was at one time essentially dependent upon rail transportation. By the early 1970s, the use of railroad tank cars had diminished to the point at which only a little over 1% of the total petroleum tonnage was hauled by the railroads. Pipeline mileage increased to become the major means of transportation.

There were two principal technological trends during that period that increased pipeline use: (1) more widespread use was made of large diameter pipe; and (2) more efficient diesel pumps became available for installation at stations along the line, replacing steam-driven models. In addition there were other developments such as (1) the introduction of welded rather than screwed-in couplings; (2) the use of high carbon steel to replace lap-welded pipe; and (3) the replacement of diesel equipment by electrically powered pumps. At the present time, the pipeline used for petroleum transportation may be up to 48 inches in diameter and may cover many thousands of miles.

Pipelines may be used to transport different types of crude oil (batch transportation). When the different batches must be kept separated to prevent mixing, slugs of kerosene, water, or occasionally inflatable rubber balls are used to separate the batches. However, there is also the possibility that the batches can be transported through the pipelines without such separators. The properties of each batch may be such that mixing, other than to form a narrow interface, is prevented. It is frequently necessary to pass cylindrical steel cleaners through the pipelines, between pumping stations, to maintain the pipeline clear of deposits.

Tank trucks are used for both lock and intermediate hauling from manufacturing and distance hauling from manufacturing and terminal points to individual domestic, commercial, and industrial consumers that maintain storage tanks on their premises. Because of costs, most bulk deliveries by truck fall within a radius of 300 miles.

Seagoing tankers (Figure 3.6), on the other hand, can be sent to any destination where a port can accommodate them and can be shifted to different routes according to need.

The seagoing tanker fleets that are owned, or used, by the world's oil companies are also responsible for the movement of a considerable portion of the world's crude oil. In fact, seagoing tankers form one of the most characteristic features associated with the transportation of petroleum. Many of these ships are of such a size that there are few ports that can handle them. Instead these large ships (VLCCs, very large crude carriers, and ULCCs, ultra large crude carriers) spend their time sailing the seas between different points, filling up and off-loading without ever entering port. Special loading jetties, artificial islands, or large buoys moored far offshore have been developed to load or off-load these tankers.

In general, the larger the tanker the lower its unit cost of transportation. As a result, the size of tankers during the 1960s and

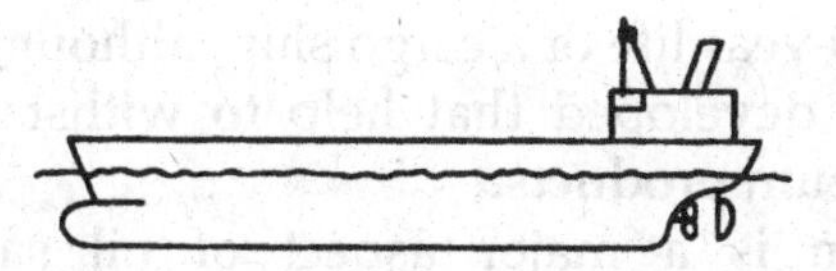

Type/cargo	Length	Beam	Draft	dwt*	Speed
1 VLCC/crude	332 m	58 m	22 m	298000	15.5 kno
2 Multi product/crude heated tanks	244 m	42 m	14.6 m	105000	14 knot
3 Clean/dirty product	183 m	32 m	12.2 m	47000	15.6 kno
4 LPG carrier	209 m	31.4 m	12.5 m	47000	—
5 LNG spherical tanks	272 m	47.2 m	11.4 m	67000	18.5 kno
6 LNG 'oblong' tanks	287 m	41.8 m	11.3 m	71470	19.2 kno

*dwt (deadweight tonnes) is the cargo capacity of the vessel

Figure 3.6 General description of sea-going vessels used for transportation of petroleum and petroleum products.

1970s grew steadily. During the 1930s and 1940s, the average size of tankers was about 12,000 deadweight ton (dwt = the number of tons of cargo, stores, and fuel that a ship can usually carry). By the 1950s a 33,000 dwt tanker was considered standard size. At present, the VLCC and ULCC classes involve tankers of over 200,000 and 300,000 dwt. A few tankers in the 500,000 dwt range also exist, and some have even exceeded this size.

The cargo space in the tankers is usually divided into two, three, or four rows of cargo tanks by longitudinal bulkheads. These are further divided into individual tanks (from 25 to 40 feet long in large vessels) by transverse bulkheads. Access to the tanks is through oil-tight hatches on the deck, cargo being loaded or discharged by means of the ship's own pumps, which may have a capacity in excess of 4000 ton/h. There has lately been some discussion about the wisdom of building such tankers with double hulls. In theory, the single-hull tanker has a better chance of staying afloat but will spill some of the cargo into the sea.

Over the past two decades it has also become evident that crude oil tankers are usually shorter lived than most other cargo ships. Crude oil cargoes can deposit corrosive sludge on the bottom of the hold, and gasoline cargoes can also have a corrosive effect on the steel of the tanks. As a result, a tanker may last only 12 years instead of the 20-year life of a cargo ship, although protective coatings have been developed that help to withstand the corrosive effects of petroleum products.

Transportation is a major aspect of oil sands exploitation (Demaison, 1977). There are four major aspects of liquid fuels production from an oil sand resource:

1. Ore recovery
2. Bitumen separation
3. Bitumen conversion to synthetic crude oil; and
4. Refining the synthetic crude oil to usable liquid fuels.

Currently, the two commercial plants carry out the first three stages on-site, and in this respect the plants are completely self-contained. The synthetic crude oil is then shipped by pipeline to a more conventional refinery site (e.g. Edmonton) for fuel upgrading to liquid fuels. However, there are constraints on the character of liquids that may be shipped by pipeline. The synthetic crude oil conveniently meets the specifications for pipeline shipment, but should an alternative means of bitumen upgrading be established, that is visbreaking (viscosity breaking), pipeline specifications must be met.

It may be deemed desirable to ship the whole bitumen feed by pipeline, in which case dilution with naphtha is necessary. The naphtha may actually be produced at the recovery site by construction of a nominal conversion operation. However, it is to be anticipated that a feed having a viscosity in excess of 15,000 cSt at 35°C (100°F) will require in excess of 0.5 barrels naphtha per barrel of bitumen. At higher temperatures, the amount of naphtha required is reduced, but light ends and even water may have to be removed to prevent undue pressure buildup in the pipeline if shipping temperatures of about 95°C (200°F) are considered.

One other aspect of transportation is the shipment of bitumen separated from tar sand (or the whole oil sand or even bitumen-enriched oil sand, produced by a less efficient once-through hot water separation) in trucks or trains. Currently, economic constraints related to the amount of material that would have to be moved to enable even a nominal conversion or upgrading operation to run continuously (hazards of weather and mechanical constraints notwithstanding) have caused these types of operation to be downgraded in priority.

Finally, it is also possible for bitumen to be emulsified and shipped by pipeline as an emulsion. This particular idea has received some attention, especially in regard to bitumen recovery by aqueous flooding methods. The idea is to produce the bitumen from the formation as an oil-in-water emulsion at the remote site followed by shipping of the emulsion to an oil recovery and upgrading site.

Seagoing vessels also transport natural gas. The gas is either transported under pressure at ambient temperatures (e.g. propane and butanes) or at atmospheric pressure but with the cargo under refrigeration (e.g. liquefied petroleum gas). For safety reasons, petroleum tankers are constructed with several independent tanks so that rupture of one tank will not necessarily drain the whole ship, unless it is a severe bow-to-stern (or stern-to-bow) rupture. Similarly, gas tankers also contain several separate tanks.

Natural gas presents different transportation requirements problems. Before World War II its use was limited by the difficulty in transporting it over long distances. The gas found in oil fields was frequently burned off; and unassociated dry gas was usually abandoned. After the war new steel alloys permitted the laying of large-diameter pipes for gas transport in the United States. The discovery of the Groningen field in the Netherlands in the early 1960s and the exploitation of huge deposits in Soviet Siberia in the 1970s and 1980s led to a similar expansion of pipelines and natural gas use in Europe.

Because of its lower density natural gas is much more expensive to ship than crude oil. Most natural gas moves by pipeline, but in the late 1960s tanker shipment of cryogenically liquefied natural gas (*LNG*) began, particularly from the producing nations in the Pacific to Japan. Special alloys are required to prevent the tanks from becoming brittle at the low temperatures (–161°C, –258°F) required to keep the gas liquid.

3.6 Products and Product Quality

Conventional crude oil is a brownish green to black liquids of specific gravity in a range from about 0.810 to 0.985 and having a boiling range from about 20°C (68°F) to above 350°C (660°F), above which active decomposition ensues when distillation is attempted. The oils contain from 0 to 35% or more of gasoline, as well as varying proportions of kerosene hydrocarbons and higher boiling constituents up to the viscous and nonvolatile compounds present in lubricant oil and in asphalt. The composition of the crude oil obtained from the well is variable and depends not only on the original composition of the oil in situ but also on the manner of production and the stage reached in the life of the well or reservoir.

For a newly opened formation and under ideal conditions the proportions of gas may be so high that the oil is, in fact, a solution of liquid in gas that leaves the reservoir rock so efficiently that a core sample will not show any obvious oil content. A general rough indication of this situation is a high ratio of gas to oil produced. This ratio may be zero for fields in which the rock pressure has been dissipated. The oil must be pumped out to as much as 50,000 ft^3 or more of gas per barrel of oil in the so-called condensate reservoirs, in which a very light crude oil (specific gravity $\leq$0.80) exists as vapor at high pressure and elevated temperature.

Thus, fluids produced from a well are seldom pure crude oil: in fact, a variety of materials may be produced by oil wells in addition to liquid and gaseous hydrocarbons. The natural gas itself may contain as impurities one or more non-hydrocarbon substances. The most abundant of these impurities is hydrogen sulfide, which imparts a noticeable odor to the gas. A small amount of this compound is considered advantageous as it gives an indication of leaks and where they occur. However, a larger amount makes the gas obnoxious and difficult to market. Such gas is referred to as sour gas and much of it is used in the manufacture of carbon black. A few natural gases contain helium, and this element does in fact occur in commercial quantities in certain gas fields; nitrogen and carbon dioxide are also found in some natural gases. Gas is usually separated at as high a pressure as possible, reducing compression costs when the gas is to be used for gaslift or delivered to a pipeline. Lighter hydrocarbons and hydrogen sulfide are removed as necessary to obtain a crude oil of suitable vapor pressure for transport yet retaining most of the natural gasoline constituents.

By far the most abundant extraneous material is water. Many wells, especially during their declining years, produce vast quantities of salt water, and disposing of it is both a serious and an expensive problem. Furthermore, the brine may be corrosive, which necessitates frequent replacement of casing, pipe, and valves, or it may be saturated so that the salts tend to precipitate upon reaching the surface. In either case, the water produced with the oil is a source of continuing trouble. Finally, if the reservoir rock is an incoherent sand or poorly cemented sandstone, large quantities of sand are produced along with the oil and gas. On its way to the surface, the sand has been known to scour its way completely through pipes and fittings.

It must also be remembered that in any field where primary production is followed by a secondary or enhanced production method, there will be noticeable differences in properties between the fluids produced (Thomas *et al.*, 1983). The differences in elemental composition may not reflect these differences to any great extent (Zou *et al.*., 1989), but more significant differences will be evident from an inspection of the physical properties. One issue that arises from the physical property data is that such oils may be outside the range of acceptability for refining techniques other than thermal options. In addition, overloading of thermal process units will increase as the proportion of the heavy oil in the refinery feedstock increases. There is a need for more and more refineries to accept larger proportions of heavy crude oils as the refinery feedstock and have the capability to process such materials.

Technologies such as alkaline flooding, microemulsion (micellar/emulsion) flooding, polymer augmented water flooding, and carbon dioxide miscible/immiscible flooding do not require or cause any change to the oil. The steaming technologies may cause some steam distillation that can augment the process when the steam distilled material moves with the steam front and acts as a solvent for oil ahead of the steam front (Pratts, 1986). Again, there is no change to the oil although there may be favorable compositional changes to the oil insofar as lighter fractions are recovered and heavier materials remain in the reservoir (Richardson *et al.*, 1992).

The technology where changes do occur involves combustion of the oil *in situ*. The concept of any combustion technology requires that the oil be partially combusted and that thermal decomposition occur to other parts of the oil. This is sufficient to cause irreversible chemical and physical changes to the oil to the extent that the product is markedly different to the oil in place. Recognition of this phenomenon is essential before combustion technologies are applied to oil recovery.

Although this improvement in properties may not appear to be too drastic, nevertheless it usually is sufficient to have major advantages for refinery operators. Any incremental increase in the units of hydrogen/carbon ratio can save amounts of costly hydrogen during upgrading. The same principles are also operative for reductions in the nitrogen, sulfur, and oxygen contents. This latter occurrence also improves catalyst life and activity as well as reduces the metals content.

In short, in situ recovery processes, although less efficient in terms of bitumen recovery relative to mining operations, may have the added benefit of leaving some of the more obnoxious constituents from the processing objective in the ground.

3.7 References

Borchardt, J.K., and Yen, T.F. 1989. Oil Field Chemistry. Symposium Series No. 396. American Chemical Society, Washington, DC.

Butler, R.M., and Mokrys, I.J. 1991. Journal of Canadian Petroleum Technology. 30(1): 97–106.

Butler, R.M., and Jiang, Q. 2000. Journal of Canadian Petroleum Technology. 39: 48–56.

Craft, B.C., and Hawkins, M.F. 1959. Applied Petroleum Reservoir Engineering. Prentice-Hall, Englewood Cliffs, New Jersey.

Demaison, G.J. 1977. The Oil Sands of Canada-Venezuela. D.A. Redford and A.G. Winestock (Editors.). Special Volume No. 17. Canadian Institute of Mining and Metallurgy. p. 9.

Forbes, R.I. 1958. A History of Technology. Oxford University Press, Oxford, England.

Frick, T.C. 1962. Petroleum Production Handbook. Volume II. McGraw-Hill, New York.

Kaufmann, R.K., and Cleveland, C.J. 2001. Oil Production in the Lower 48 States; Economic, Geological, and Institutional Determinants. The Energy Journal. 22: 27–49.

Mokhatab, S., Poe, W.A., and Speight, J.G. 2006. Handbook of Natural Gas Transmission and Processing. Elsevier, Amsterdam, The Netherlands.

Pratts, M. 1986. Thermal Recovery. Society of Petroleum Engineers, New York. Volume 7.

Ranney, M.W. 1979. Crude Oil Drilling Fluids. Noyes Data Corp., Park Ridge, New Jersey.

Richardson, W.C., Fontaine, M.F., and Haynes, S. 1992. Paper No. SPE 24033. Western Regional Meeting. Bakersfield, California. March 30-April 1.

Speight, J.G. 2000. The Desulfurization of Heavy Oils and Residua. 2nd Edition. Marcel Dekker Inc., New York.

Speight, J.G. 2008. Handbook of Synthetic Fuels. McGraw-Hill, New York.

Speight, J.G. 2009. Enhanced Recovery Methods for Heavy Oil and Tar Sands, Gulf Publishing Company, Houston, Texas.

Thomas, K.P., Barbour, R.V., Branthaver, J.F., and Dorrence, S.M. 1983. Fuel 62: 438.

United States Patent 5,407,009. 1995. Butler, R.M., and Mokrys, I.J. Process and apparatus for the recovery of hydrocarbons from a hydrocarbon deposit. April 18.

United States Patent 5,607,016. 1995. Butler, R.M., and Mokrys, I.J. Process and apparatus for the recovery of hydrocarbons from a hydrocarbon deposit. March 4.

Zou, J., Gray, M.R., and Thiel, J. 1989. AOSTRA J. Res. 5: 75.

4

Crude Oil Classification and Benchmarks

Recent fluctuations in the price of oil have triggered the debate regarding the level of world oil reserves, and the capacity to meet future energy demand has taken on a new impetus. This has led to reinvestigation of the methods of crude oil classification and classification of reserves.

For the purpose of the book, petroleum is a naturally occurring mixture of hydrocarbons, generally in a liquid state, that may also include compounds of sulfur nitrogen oxygen metals and other elements (ASTM D4175), which occurs in sedimentary rock deposits throughout the world (Speight, 2007). However, the definition of petroleum-associated materials has been varied, unsystematic, diverse, and often archaic and it is only recently that some attempt has been made to define these materials in a meaningful manner (Chapter 1). Thus, it is not surprising that attempts to classify petroleum have also evolved, and it is the purpose of this chapter to review these methods and present them for further consideration in terms of pricing strategies.

Crude oil is the primary input into the petroleum refining industry. Therefore, basic information about the production and distribution of crude oil will provide useful context for the downstream business. However, regardless of the source of crude oil, the price is determined in the world market and both imported and domestic crude oil is priced according to the supply/demand balance and pricing dynamics on the world oil market. In this respect, many refiners have very little influence on the price they pay for crude oil.

The overall economics or viability of a refinery depends on the interaction of three key elements: the choice of crude oil used (crude slates), the complexity of the refining equipment (refinery configuration) and the desired type and quality of products produced (product slate). Refinery utilization rates and environmental considerations also influence refinery economics.

Using more expensive, lighter and sweeter crude oil requires less refinery upgrading, but supplies of light, sweet crude oil are decreasing and the differential between heavier and sourer crude oils is increasing. Using cheaper, heavier crude oil means more investment in upgrading processes. Costs and payback periods for refinery processing units must be weighed against anticipated crude oil costs and the projected differential between light and heavy crude oil prices.

Crude slates and refinery configurations must take into account the type of products that will ultimately be needed in the marketplace. The quality specifications of the final products are also increasingly important as environmental requirements become more stringent.

Another aspect of crude oil pricing technology that is often ignored, especially when dealing with crude oil pricing, is the Hubbert peak theory.

The Hubbert peak theory, also known as peak oil, postulates that future petroleum production, whether for individual oil wells, entire oil fields, whole countries, or worldwide production, will eventually peak and then decline at a similar rate to the rate of increase before the peak as these reserves are exhausted. The theory also suggests a method to calculate the timing of this peak, based on past production rates, the observed peak of past discovery rates, and proven oil reserves.

The theory arose in 1956 when M. King Hubbert correctly predicted US oil production would peak around 1971. When this occurred and the US began losing its excess production capacity, OPEC gained the ability to manipulate oil prices, leading to the 1973 and 1979 oil crises.

However, controversy surrounds predictions of the timing of the global peak, as these predictions are dependent on the past production and discovery data used in the calculation as well as how unconventional reserves are considered. Supergiant fields have been discovered in the past two decades, such as Azadegan, Carioca/Sugar Loaf, Tupi, Jupiter, Ferdows/Mounds/Zagheh, Tahe, Jidong Nanpu/Bohai Bay, West Kamchatka, and Kashagan, as well as tremendous reservoir growth from places such as the Bakken and massive syncrude operations in Venezuela and Canada. While past understanding of total oil reserves changed with newer scientific understanding of petroleum geology, current estimates of total oil reserves have been in general agreement since the 1960s. Furthermore, predictions regarding the timing of the peak are highly dependent on the past production and discovery data used in the calculation.

It is difficult to predict the oil peak in any given region, due to the lack of transparency in accounting of global oil reserves. Based on available production data, proponents have previously predicted the peak for the world to be in years 1989, 1995, or 1995–2000. Some of these predictions date from before the recession of the early 1980s, and the consequent reduction in global consumption, the effect of which was to delay the date of any peak by several years. Just as the 1971 U.S. peak in oil production was only clearly recognized after the fact, a peak in world production will be difficult to discern until production clearly drops off.

Clearly such concerns, whether real or truly theoretical, cannot help but have some influence on the price of oil.

Thus, the classification of crude oil and crude oil reserves is an important indicator for the value of an oil company and the netback from a barrel of oil (i.e. the linkage between the price of curd oil and the products derived from it) ultimately determines the intrinsic value of a company. Because largely its reserves and resources drive the value of a company, it is one of the main drivers of the market capitalization of listed companies.

4.1 Crude Oil Classification

All crude oil is not valued equally because it varies markedly in appearance and properties, particularly in ease of recovery from the reservoir and production of saleable products in the refinery, depending on its composition. It is usually black or dark brown, although it may be yellowish or even greenish. In fact, the need to classify crude oil arose because the quality of crude oil dictates the level of processing and conversion necessary to achieve what a refiner sees as an optimal mix of products.

Different types of crude oil yield a different mix of products, depending on the crude oil's natural qualities. Their density, measured as American Petroleum Institute (API) gravity, and their sulfur content, typically differentiates crude oil types. Crude oil with a low API gravity is considered a heavy crude oil and typically has higher sulfur content and a larger yield of lower-valued products. Therefore, the lower the API of a crude oil, the lower the value it has to a refiner as it will either require more processing or yield a higher percentage of lower-valued products such as heavy fuel oil, which can often sell for less than crude oil.

Using more expensive, lighter and sweeter crude oil requires less refinery upgrading. However, supplies of light, sweet crude oil are decreasing and the differential between heavier and sourer crudes is increasing. Using cheaper heavier crude oil means more investment in upgrading processes. Costs and payback periods for refinery processing units must be weighed against anticipated crude oil costs and the projected differential between light and heavy crude oil prices. Furthermore, the difference in value between light and heavy oil is primarily determined in the market for each type and a widening of the differential generally leads to poorer profitability for heavy oil producers.

Crude slates and refinery configurations must take into account the type of products that will ultimately be needed in the marketplace. The quality specifications of the final products are also increasingly important as environmental requirements become more stringent.

The original methods of classification arose because of the commercial interest in and recognition of the different types of crude oil, and were a means of providing refinery operators with a rough guide to processing conditions. It is therefore not surprising that systems based on a superficial inspection of a physical property, such as specific gravity or API (Baumé) gravity, are easily applied, and are actually used to a large extent in expressing the quality of crude oils. Such a system is approximately indicative of the general character of a crude oil as long as materials of one general type are under consideration. For example, among crude oils from a particular area, an oil of 40°API (specific gravity = 0.825) is usually more valuable than one of 20°API (specific gravity = 0.934) because it contains more light fractions (e.g., gasoline) and fewer heavy, undesirable asphaltic constituents.

Crude oil is a mixture of hydrocarbons, often found together with natural gas. The main characteristics of by which crude oil is defined are: (1) density which, in the oil industry, is usually measured by its API gravity, and (2) sulfur content. Thus, crude oil is usually described as sweet (low sulfur) or sour (high sulfur) and light or heavy, depending on its density. Heavier oils may also be described as medium and extra heavy oil although the definition of this later is open to question. Although the classification of crude oil may be defined by API gravity and sulfur content, this is not the complete story and other methods of classification are available.

Only those methods of classification that have the potential to affect the price are included here. Other methods are available (Speight, 2007) but are more specific to the scientific aspect of crude oil rather than to the pricing and are not included here.

4.1.1 Classification as a Hydrocarbon Resource

Petroleum is referred to generically as a fossil energy resource and is further classified as a hydrocarbon resource and, for illustrative or comparative purposes in this text, coal and oil shale kerogen have also been included in this classification. However, the inclusion of coal and oil shale under the broad classification of hydrocarbon resources has incorrectly required that the term hydrocarbon be expanded to include the macromolecular non-hydrocarbon species that constitute coal and oil shale kerogen. The use of the term organic sediments would be more accurate (Figure 4.1). The inclusion of coal and oil shale kerogen in the category hydrocarbon resources is due to the fact that these two natural resources will produce hydrocarbons on high-temperature processing. Therefore, if coal and oil shale kerogen are to be included in the term hydrocarbon resources, it is more appropriate that they be classed as hydrocarbon-producing resources under the general classification of organic sediments (Figure 4.2).

4.1.2 Classification by Chemical Composition

Composition refers to the specific mixture of chemical compounds that constitute petroleum.

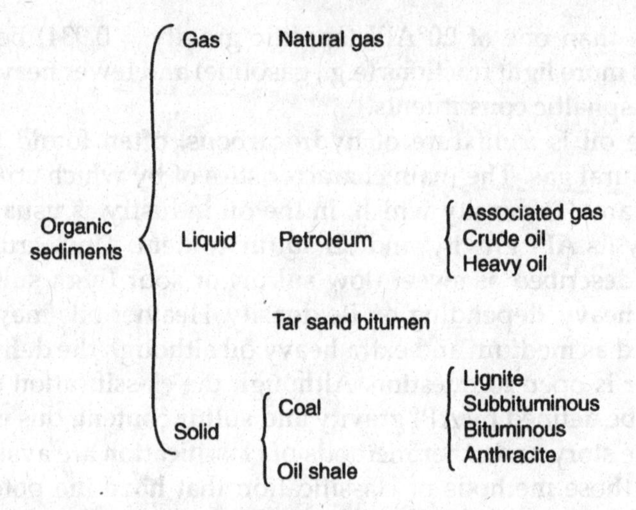

Figure 4.1 Subdivision of the earth's organic sediments.

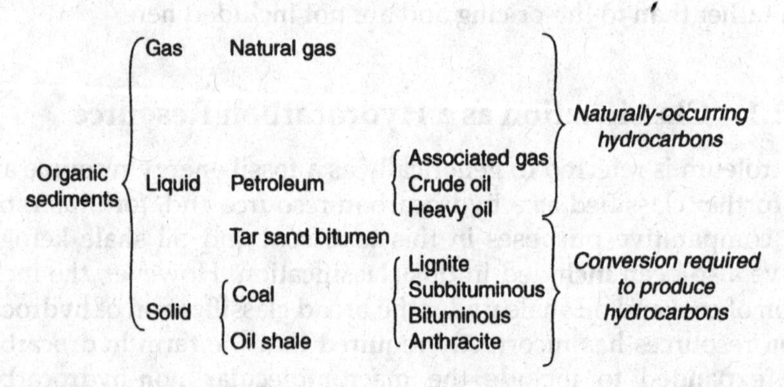

Figure 4.2 Classification of the earth's organic sediments according to hydrocarbon occurrence and production.

The composition of petroleum obtained from the well is variable and depends not only on the original composition of the oil in situ but also on the manner of production and the stage reached in the life of the well or reservoir. In fact, petroleum (conventional crude oil) ranges from a brownish green to black liquid having a specific gravity (at 60°F, 15.6°C) that varies from about 0.75 to 1.00 (57 to 10°API), with the specific gravity of most crude oils falling in the range 0.80 to 0.95 (45 to 17°API). The boiling range of petroleum varies from about 20°C (68°F) to above 350°C (660°F), above which active decomposition ensues when distillation is attempted.

Furthermore, petroleum varies in composition from one oil field to another, from one well to another in the same field, and even from one level to another in the same well. This variation can be in both molecular weight and the types of molecules present in petroleum.

4.1.3 Density and API Gravity

The use of density values has been advocated for quantitative application using a scheme based on the American Petroleum Institute (API) gravity, which offers a wider scale of values by which crude oil can be classified and priced. As indicated above, the density of crude oil varies slightly from conventional crude to heavy crude.

Light crude oil is usually oil having API gravity above 30 or 35°API. Medium crude oil falls into the API gravity range is below these numbers. Heavy oil is considered to be those petroleum-type materials that had gravity somewhat less than 20°API, with the heavy oils falling into the API gravity range 10 to 15° (e.g., Cold Lake crude oil = 12°API) and bitumen falling into the 5 to 10°API range (e.g., Athabasca bitumen = 8°API).

However, the assignment of specific numbers to the classification of petroleum is fraught with difficulty. Using such points of demarcation does not circumvent the question that must arise when one considers a material having API gravity equal to 9.9 and one material having API gravity equal to 10.1; nor does the point of demarcation make allowance for the limitations of the accuracy of the analytical method. The use of one physical parameter, be it API gravity or any other physical property for that matter, is inadequate to the task of classifying conventional petroleum, heavy oil, and tar sand bitumen.

4.1.4 Viscosity

At the same time, in concert with the use of API gravity, the line of demarcation between petroleum and heavy oil vis-à-vis tar sand bitumen has been drawn at 10,000 centipoises. Briefly, materials having a viscosity of less than 10,000 centipoises are conventional petroleum and heavy oil, while tar sand bitumen has a viscosity greater than 10,000 centipoises. The use of such a scale requires a fine line of demarcation between the various crude oils, heavy oils, and bitumen to the point where it would be confusing to have to differentiate between a material having a viscosity of 9950 cp and one having a viscosity of 10,050 cp. Furthermore, the inaccuracies (i.e., the limits of experimental error) of the method of measuring viscosity also increase the potential for misclassification.

The use of viscosity also does not circumvent the use of one physical property and the difference between a material having viscosity equal to 49,900 and 50,100 centipoises (or 99,900 and 100,100 centipoises).

4.1.5 UOP Characterization Factor

This factor is perhaps one of the more widely used derived characterization or classification factors and is defined by the formula

$$K = \sqrt[3]{T_B}/d$$

T_B is the average boiling point in degrees Rankine (°F + 460) and d is the specific gravity 60°/60°F. This factor has been shown to be additive on a weight basis. It was originally devised to show the thermal cracking characteristics of heavy oils; thus, highly paraffin oils have K in the range 12.5 to 13.0 and cyclic (naphthene) oils have K in the range 10.5 to 12.5.

4.1.6 Pour Point

The pour point of petroleum or a petroleum product is the lowest temperature at which oil will move, pour, or flow when it is chilled without disturbance under definite conditions (ASTM D97). In fact, the pour point of oil, when used in conjunction with the reservoir temperature, gives a better indication of the condition of the oil in the reservoir than the viscosity. Thus, the pour point and reservoir temperature present a more accurate assessment of the condition of the oil in the reservoir, being an indicator of the mobility of the oil in the reservoir. When used in conjunction with reservoir temperature, the pour point gives an indication of the liquidity of the heavy oil or bitumen and, therefore, the ability of the heavy oil or bitumen to flow under reservoir conditions. In summary, the pour point is an important consideration because, for efficient production, additional energy must be supplied to the reservoir by a thermal process to increase the reservoir temperature beyond the pour point.

A classification method that uses the pour point of the oil and the reservoir temperature adds a specific qualification to the term extremely viscous as it occurs in the definition of tar sand. In fact, when used in conjunction with the recovery method, pour point offers more general applicability to the conditions of the oil in the reservoir or the bitumen in the deposit, and a comparison of the two temperatures shows promise and warrants further consideration.

4.1.7 Recovery Method

All of the classification systems described above are based on the assumption that petroleum can be more or less characterized by the properties of one or a few fractions. However, the properties of certain fractions of a crude oil are not always reflected in those of other fractions of the same oil; plus, any method of classification in which the properties of a certain fraction are extrapolated to the whole crude oil must be applied with caution as serious errors can arise.

In order to classify petroleum, heavy oil, and tar sand bitumen, the use of a single parameter such as viscosity is not enough (Speight, 2007). Other properties, such as API gravity, elemental analysis, composition, the properties of the fluid in the reservoir, and the method of recovery need to be acknowledged.

The United States Congress has defined tar sands as "the several rock types that contain an extremely viscous hydrocarbon which is not recoverable in its natural state by conventional oil well production methods including currently used enhanced recovery techniques" (US Congress, 1976). By inferences, conventional petroleum and heavy oil can be recovered by currently used enhanced recovery techniques. Furthermore, conventional petroleum can be recovered by primary and secondary techniques.

However, the methods of pricing crude oil require something more than these general properties and methods of classification, although some of the parameters, such as API gravity and sulfur content, are still employed. It is here that the need for benchmark crude oils enters the arena.

4.2 Classification of Reserves

The classification of reserves and, hence, reserve estimates for crude oil, are very difficult to develop. There are various versions of crude oil reserves of which the most frequently used is the BP Statistical Review of World Energy (BP, 2008). However, even the best laid plans and numbers are subject to debate. As a result, the classification of reserves is discussed here so that the reader may compare the recognized systems of reserve classification and wonder why all oil-producing countries do not subscribe to similar systems or develop their own systems. In fact, the reserves estimates of various oil-producing nations have already come under question (Campbell and Laherrère, 1998; Aluko, 2004).

Reporting data on oil production or oil reserves is either a technical or a political act. The Society of Petroleum Engineers classify crude oil reserves to satisfy the technical needs, while the Securities Exchange Commission (SEC) classify reserves to satisfy bankers and shareholders and requires oil companies listed on the US stock market to report only proved reserves and to omit probable reserves that are reported in the rest of the world.

Various definitions have been applied to energy reserves, but the crux of the matter is the amount of a resource that is recoverable using current technology. In fact, the definitions that are used to describe petroleum reserves are often misunderstood because they are not adequately defined at the time of use. Therefore, as a means of alleviating this problem, it is pertinent to consider the definitions used to describe the amount of petroleum that remains in subterranean reservoirs.

Petroleum is a resource; in particular, petroleum is a fossil fuel resource. A resource is the entire commodity that exists in the sediments and strata, whereas the reserves represent that fraction of a commodity that can be recovered economically.

Reserves are those quantities of petroleum claimed to be commercially recoverable by application of development projects to known accumulations under defined conditions. Reserves must satisfy four criteria:

1. They must be discovered through one or more exploratory wells
2. They must be recoverable using existing technology
3. They must be commercially viable
4. They must be remaining in the ground.

As already noted (Chapter 2) one of the basic requirements of reservoir studies is to provide certifications of oil and gas reserves and resources, which is known as *reservoir evaluation*. Crude oil reserves are the estimated quantities of crude oil that are claimed to be recoverable under existing economic and operating conditions.

Geological and geochemical analyses of hydrocarbons and organic matter in source rocks are critical in hydrocarbon exploration and in examining the extent and continuity of hydrocarbon reservoirs. Hydrocarbons may be correlated with source rock to verify the results of basin modeling.

The estimation of crude oil reserves (Chapter 1) is usually subdivided into two components: (1) the total reserves contained in the reservoir — usually termed the resource or original-oil-in-place (OOIP) and (2) recoverable reserves — that portion of the resource, as of the date the reserves are calculated, is economically efficient given market conditions and rational use of modern extraction equipment and technologies. The reserve estimate will change with the price of oil and with the evolution of new and more efficient recovery technologies. Thus, early estimates of the reserves of an oil field may appear to be conservative insofar as the number increases with time; this phenomenon is referred to as reserves growth.

The term oil-in-place (OIP) may also be used for the resource, but it is more appropriate to indicate the oil remaining after recovery has commenced and been operating for a time.

The ratio of producible oil reserves to total oil in place for a given field is often referred to as the recovery factor, which varies from oil field to oil field. In fact, the recovery factor of any particular field may change over time, based on operating history and in response to changes in technology and economics.

Many oil-producing nations do not reveal their reservoir engineering field data, but instead provide unaudited claims for their oil reserves.

The United States had two main ways of announcing reserves: (1) the standards developed by the Society of Petroleum Engineers (SPE) and (2) the standards employed by the Securities and Exchange Commission (SEC). Another system of standards, developed by Russia, has also become available recently, and, because of Russia being an oil-exporting country, is also noteworthy.

However, the use of the term reserves as being descriptive of the resource is subject to much speculation. In fact, it is subject to word variations; for example, reserves are classed as proved, unproved, probable, possible, and undiscovered — each with its own particular definition.

4.2.1 SPE Standards

The standards developed by the Society of Petroleum Engineers take into account not only the probability that hydrocarbons are physically present in a given geological formation, but also the economic viability of recovering the reserves. This includes many factors, such as exploration and drilling costs, ongoing production costs, transportation costs, taxes, prevailing prices for the products, and other factors that influence the economic viability of a given deposit.

Within these standards, reserves are classified as proved, probable and possible, based on both geological and commercial factors.

All reserve estimates involve uncertainty, depending on the amount of reliable geologic and engineering data available and the interpretation of those data. The relative degree of uncertainty can be expressed by dividing reserves into two principal classifications — proven (proved) and unproven (unproved). Unproven reserves can further be divided into two subcategories — probable and possible — to indicate the relative degree of uncertainty about their existence.

Proved reserves are reserves of petroleum that are confirmed with a high degree of certainty, and are actually found by drilling operations and are recoverable by means of current technology. They have a high degree of accuracy and are frequently updated as the recovery operation proceeds. Proven reserves have a reasonable certainty (normally at least 90% confidence) of being recoverable under existing economic and political conditions, and using existing technology. Industry specialists refer to this as P90 (i.e. having a 90% certainty of being produced). Proved reserves are also known in the industry as 1P. They may be updated by means of reservoir characteristics, such as production data, pressure transient analysis, and reservoir modeling.

Probable reserves are those reserves of petroleum that are nearly certain but about which a slight doubt exists. Possible reserves are those reserves of petroleum with an even greater degree of uncertainty about recovery, but about which there is some information. An additional term, potential reserves, is also used on occasion; these reserves are based upon geological information about the types of sediments where such resources are likely to occur and they are considered to represent an educated guess. Then, there are the so-called undiscovered reserves, which are little more than figments of the imagination. The terms undiscovered reserves or undiscovered resources should be used with caution, especially when applied as a means of estimating reserves of petroleum reserves. The data is very speculative and regarded by many energy scientists as having little value other than unbridled optimism.

Probable reserves are those reserves of petroleum that are nearly certain but about which a slight doubt exists. Probable reserves are based on median estimates, and claim a 50% confidence level of recovery. Industry specialists refer to this as P50 (i.e. having a 50% certainty of being produced), which is also referred to in the industry as 2P (proved plus probable).

Possible reserves are those reserves of petroleum with an even greater degree of uncertainty about recovery but about which there is some information. Possible reserves have a less likely chance of being recovered than probable reserves. This term is often used for reserves that are claimed to have at least a 10% certainty of being produced (P10). Reasons for classifying reserves as possible include varying interpretations of geology, reserves not producible at commercial rates, uncertainty due to reserve infill (seepage from adjacent areas), and projected reserves based on future recovery methods. This is referred to in the industry as 3P (proved plus probable plus possible).

An evaluation of proved, probable, and possible natural gas reserves naturally involves multiple uncertainties. The accuracy of any reserves evaluation depends on the quality of available information and engineering and geological interpretation. Based on the results of drilling, testing, and production after the audit date, reserves may be significantly restated upwards or downwards. Changes in the price of natural gas, gas condensate or crude oil may also affect our proved and probable reserves estimates, as well as estimates of its future net revenues and net present worth. This is because the reserves are evaluated, and the future net revenues and net present worth are estimated based on prices and costs as of the audit date.

Recently, an additional term — potential reserves — is also used on occasion; these reserves are based upon geological information about the types of sediments where such resources are likely to occur and they are considered to represent an educated guess. The term inferred reserves is also commonly used in addition to, or in place of, potential reserves. Inferred reserves are regarded as of a higher degree of accuracy than potential reserves, and the term is applied to those reserves that are estimated using an improved understanding of reservoir frameworks. The term also usually includes those reserves that can be recovered by further development of recovery technologies. Oil companies and government agencies use unproved reserves internally for future planning purposes.

More recently, the Society for Petroleum Engineers has developed a resource classification system (Figure 1.1) that moves away from systems in which all quantities of petroleum that are estimated to be initially-in-place are used. Some users consider only the estimated recoverable portion to constitute a resource. In these definitions, the quantities estimated to be initially in place are (1) total petroleum-initially-in-place, (2) discovered petroleum-initially-in-place, and (3) undiscovered petroleum-initially-in-place. The recoverable portions of petroleum are defined separately as (1) reserves, (2) contingent resources, and (3) prospective resources. In any case, reserves are a subset of resources and are those quantities petroleum that are discovered (i.e. in known accumulations), recoverable, commercial and remaining.

The total petroleum initially in place is that quantity of petroleum that is estimated to exist originally in naturally occurring accumulations. The total petroleum initially in place is, therefore, that quantity of petroleum that is estimated to be contained in known accumulations, plus those quantities already produced therefrom and those estimated quantities in accumulations yet to be discovered. The total petroleum initially in place may be subdivided into discovered petroleum initially in place and undiscovered petroleum initially in place, with discovered petroleum initially in place being limited to known accumulations.

It is recognized that the quantity of petroleum initially in place may constitute potentially recoverable resources because the estimation of the proportion that may be recoverable can be subject to significant uncertainty and will change with variations in commercial circumstances, technological developments, and data availability. A portion of those quantities classified as unrecoverable may become recoverable resources in the future as commercial circumstances change, technological developments occur, or additional data are acquired.

Discovered petroleum initially in place may be subdivided into commercial and sub-commercial categories, with the estimated potentially recoverable portion being classified as reserves and contingent resources respectively, as defined below.

Reserves are those quantities of petroleum which are anticipated to be commercially recovered from known accumulations from a given date forward (http://www.spe.org/spe/jsp/basic/0,,1104_1718,00.html).

Estimated recoverable quantities from known accumulations that do not fulfill the requirement of commerciality should be classified as contingent resources, as defined below. The definition of commerciality for an accumulation will vary according to local conditions and circumstances and is left to the discretion of the country or company concerned. However, reserves must still be categorized according to specific criteria, therefore proved reserves will be limited to those quantities that are commercial under current economic conditions, while probable and possible reserves may be based on future economic conditions. In general, quantities should not be classified as reserves unless there is an expectation that the accumulation will be developed and placed on production within a reasonable timeframe.

In certain circumstances, reserves may be assigned even though development may not occur for some time. An example of this would be where fields are dedicated to a long-term supply contract and will only be developed as and when they are required to satisfy that contract.

Contingent resources are those quantities of petroleum that are estimated, on a given date, to be potentially recoverable from known accumulations, but which are not currently considered as commercially recoverable. Some ambiguity may exist between the definitions of contingent resources and unproved reserves. This is a reflection of variations in current industry practice, but if the degree of commitment is not such that the accumulation is expected to be developed and placed on production within a reasonable timeframe, the estimated recoverable volumes for the accumulation be classified as contingent resources. Contingent resources may include, for example, accumulations for which there is currently no viable market, or where commercial recovery is dependent on the development of new technology, or where evaluation of the accumulation is still at an early stage.

Undiscovered petroleum initially in place is that quantity of petroleum that is estimated, on a given date, to be contained in accumulations that are yet to be discovered. The estimated potentially recoverable portion of undiscovered petroleum initially in place is classified as prospective resources, which are those quantities of petroleum that are estimated, on a given date, to be potentially recoverable from undiscovered accumulations.

The ultimate recoverable resource (URR) is the total quantity of oil that will ever be produced, including the nearly 1 trillion barrels extracted to date. Estimated ultimate recovery (EUR) is the quantity of petroleum that is estimated, on a given date, to be potentially recoverable from an accumulation, plus those quantities already produced therefrom. EUR is not a resource category but it is a term that may be applied to an individual accumulation of any status/maturity, whether discovered or undiscovered.

Petroleum quantities classified as reserves, contingent resources or prospective resources should not be aggregated with each other without due consideration of the significant differences in the criteria associated with their classification. In particular, there may be a significant risk that accumulations containing contingent resources or prospective resources will not achieve commercial production.

The range of uncertainty reflects a reasonable range of estimated potentially recoverable volumes for an individual accumulation. Any estimation of resource quantities for an accumulation is subject to both technical and commercial uncertainties, and should, in general, be quoted as a range. In the case of reserves, and where appropriate, this range of uncertainty can be reflected in estimates for proved reserves (1P), proved plus probable reserves (2P) and proved plus probable plus possible reserves (3P) scenarios. For other resource categories, the terms low estimate, best estimate, and high estimate are recommended.

The term best estimate is used as a generic expression for the estimate considered being the closest to the quantity that will actually be recovered from the accumulation between the date of the estimate and the time of abandonment. If probabilistic methods are used, this term would generally be a measure of central tendency of the uncertainty distribution. The terms low estimate and high estimate should provide a reasonable assessment of the range of uncertainty in the best estimate.

For undiscovered accumulations, prospective resources, the range will, in general, be substantially greater than the ranges for discovered accumulations. However, in all cases, the actual range will be dependent on the amount and quality of data (both technical and commercial) that is available for that accumulation. As more data becomes available for a specific accumulation (e.g. additional wells, reservoir performance data), the range of uncertainty in the estimated ultimate recovery for that accumulation should be reduced.

The low estimate, best estimate, and high estimate of potentially recoverable volumes should reflect some comparability with the reserves categories of proved reserves, proved plus probable reserves, and proved plus probable plus possible reserves, respectively. While there may be a significant risk that sub-commercial or undiscovered accumulations will not achieve commercial production, it is useful to consider the range of potentially recoverable volumes independently of such a risk.

At some time in the future, resources may become reserves as a result of improvements in recovery techniques which may either make the resource accessible or bring about a lowering of the recovery costs and render winning of the resource an economical proposition. In addition, other uses may also be found for a commodity, and the increased demand may result in an increase in price. Alternatively, a large deposit may become exhausted and unable to produce any more of the resource thus forcing production to focus on a resource that is lower grade but has a higher recovery cost.

Finally, peak oil is not the phenomenon of running out of oil. It is the subsequent decline of the production rate of oil and is, in fact, the point in time when the maximum rate of global crude oil recovery is reached, after which the rate of production enters a continuous decline phase. The concept is based on the observed production rates of individual oil wells, and the combined production rate of a field of related oil wells. The aggregate production rate from an oil field over time appears to grow exponentially until the rate peaks and then declines, sometimes rapidly, until the field is depleted. It has been shown to be applicable to the sum of a nation's domestic production rate, and is similarly applied to the global rate of petroleum production.

4.2.2 SEC Standards

The standards set by the Securities and Exchange Commission (SEC) in the United States differ in certain material respects from SPE standards. The principal technical difference is related to the certainty of existence (i.e. proved or proven reserves).

Under the SPE standards, reserves in undeveloped drilling sites that are located more than one well location from a commercial producing well may be classified as proved reserves if there is reasonable certainty that they exist. However, under SEC standards, the existence of the reserves must be demonstrated with certainty before they may be classified as proved reserves.

Proved reserves are the only type of reserves that the Securities and Exchange Commission allows oil companies to report to investors. Companies listed on U.S. stock exchanges must substantiate their claims, but many governments and national oil companies do not disclose verifying data to support their claims. The practice of reporting only proved reserves can lead to a strong reserve growth, as 90% of the annual reserves oil addition come from revisions of old fields, showing that the assessment of the fields was poorly reported. This reserve growth of conventional oil reserves is often (sometime wrongly) attributed to technological progress. Technical data, on which development decisions are taken, exist but they are generally held to be confidential and not for general dissemination.

Furthermore, under the SEC standards, oil in reservoirs may not be classified as proved reserves if they will be recovered after the expiration of a current license period unless the license holder has the right to renew the license and there is a demonstrated history of license renewal. However, under the SPE standards, proved reserves are projected to the economic production life of the evaluated fields.

Accordingly, information relating to estimated proved crude oil reserves under SEC standards is not necessarily indicative of information that would be reported under SEC standards in an offering document registered with the SEC. In addition, SEC standards do not permit the presentation of reserves other than proved reserves.

Under the SEC regulations, proved oil are the estimated quantities of crude oil, which geological and engineering data demonstrate with reasonable certainty to be recoverable in future years from known reservoirs under existing economic and operating conditions, including prices and costs as of the date the estimate is made. Furthermore, reservoirs are considered proved if economic producibility is supported by either actual production data or a conclusive formation test.

Reserves that can be produced economically through application of improved recovery techniques, such as fluid injection, are included in the proved classification when successful testing by a pilot project, or the operation of an installed program in the reservoir, provides support for the engineering analysis on which the project or program was based. However, estimates of proved reserves do not include the following:

1. Oil that may become available from known reservoirs but is classified separately as indicated additional reserves
2. Crude oil where the recovery subject to reasonable doubt because of uncertainty as to geology, reservoir characteristics, or economic factors
3. Crude oil that may occur in undrilled prospects, and
4. Crude oil, natural gas, and natural gas liquids, that may be recovered from oil sale, coal, gilsonite, and other such sources.

Proved reserves are further subdivided into proved developed (PD) and proved undeveloped (PUD). PD reserves are reserves that can be produced with existing wells and perforations, or from additional reservoirs where minimal additional investment (operating expense) is required. Additional oil and gas expected to be obtained through the application of fluid injection or other improved recovery techniques for supplementing the natural forces and mechanisms of primary recovery can be included as proved developed reserves only after testing by a pilot project or after the operation of an installed program has confirmed through production response that increased recovery will be achieved.

Proved undeveloped reserves are reserves that will be recovered from new wells on undrilled acreage, or from existing wells where a relatively major expenditure is required for recompletion. Reserves on undrilled acreage are limited to those drilling units offsetting productive units that are reasonably certain of production when drilled. Proved reserves for other undrilled units can be claimed only where it can be demonstrated with certainty that there is continuity of production from the existing productive formation. Estimates for proved undeveloped reserves are not allowed for any acreage for which an application of fluid injection or other improved recovery technique is contemplated, unless such techniques have been proved effective by actual tests in the area and in the same reservoir.

4.2.3 Russian Standards

The Russian classification of crude oil reserves differs significantly from SEC standards and SPE standards with respect to the manner and extent to which commercial factors are taken into account in calculating reserves. While the Russian reserves system focuses on the actual physical presence of hydrocarbons in geological formations, the reserves of crude oil are estimated based on the probability of such physical presence (http://econ.la.psu.edu/CAPCP/RussianData/data/classification.pdf).

The system is based solely on the analysis of geological attributes. Explored reserves are represented by categories A, B, and C1; preliminary estimated reserves are represented by category C2; potential resources are represented by category C3; and forecasted resources are represented by categories D1 and D2.

Category A reserves are calculated on the part of a deposit drilled in accordance with an approved development project for the oil or natural gas field and are reserves that have been analyzed in sufficient detail to define the following comprehensively:

1. The type, shape and size of the reservoir
2. The level of hydrocarbon saturation
3. The reservoir type; the nature of changes in the reservoir characteristics
4. The hydrocarbon saturation of the productive strata of the deposit
5. The content and characteristics of the hydrocarbons
6. The major features of the deposit that determine the conditions of its development (mode of operations, well productivity, strata pressure, natural gas, gas condensate and crude oil balance).

Category B reserves represents the crude oil which has been determined on the basis of commercial flows of oil in wells. The classification of reserves into this category include the following:

1. The shape and size of the reservoir
2. The oil saturation depth and type of the reservoir
3. The nature of changes in the reservoir characteristics
4. The oil saturation of the productive strata of the reservoir
5. The composition and characteristics of the crude oil.

Category C1 reserves are the reserves of a reservoir (or of a portion thereof) where the oil content has been determined on the basis of commercial flows of oil and positive results of geological and geophysical exploration of non-probed wells. In addition, the type, shape, and size of the reservoir as well as the formation structure of the oil-bearing reservoirs have been determined from the results of drilling exploration and production wells and by those geological and geophysical exploration techniques that have been field-tested for the applicable area. The lithology (rock type and content), reservoir type and characteristics, oil and natural gas saturation, oil displacement ratio, and effective oil and natural gas saturation depth of the productive strata should have been studied based on drill cores and geophysical well exploration materials. The composition and characteristics of crude oil and well productivity are also necessary for this category.

Category C1 reserves are estimated on the basis of results of geological exploration work and production drilling and must have been studied in sufficient detail to yield data from which to draw up either a trial industrial development project in the case of a natural gas field or a technological development scheme in the case of an oil field.

Category C2 reserves are preliminary estimated reserves of a deposit calculated on the basis of geological and geophysical research of unexplored sections of deposits adjoining sections of a field containing reserves of higher categories and of untested deposits of explored fields. The shape, size, structure, level, reservoir types, content, and characteristics of the hydrocarbon deposit are determined in general terms based on the results of the geological and geophysical exploration and information on the more fully explored portions of a deposit. Category C2 reserves are used to determine the development potential of a field and to plan geological, exploration and production activities.

Category C3 resources are prospective reserves prepared for the drilling of (1) traps within the oil-and-gas bearing area, delineated by geological and geophysical exploration methods tested for such area and (2) the formation of explored fields which have not yet been exposed by drilling. The form, size, and stratification conditions of the assumed deposit are estimated from the results of geological and geophysical research. The thickness, the reservoir characteristics of the formations, and the composition and characteristics of hydrocarbons are assumed to be analogous to those for explored fields. Category C3 resources are used in the planning of prospecting and exploration work in areas known to contain other reserve bearing fields.

Category D1 resources are calculated based on the results the region's geological, geophysical and geochemical research and by analogy with explored fields within the region being evaluated. Category D1 resources are reserves in lithological and stratigraphic series that are evaluated within the boundaries of large regional structures confirmed to contain commercial reserves of oil and natural gas.

Category D2 resources are calculated using assumed parameters on the basis of general geological concepts and by analogy with other, better studied regions with explored oil and natural gas fields. Category D2 resources are reserves in lithological and stratigraphic series that are evaluated within the boundaries of large regional structures not yet confirmed to contain commercial reserves of oil and natural gas. The prospects for these series that prove to be oil- and gas-bearing are evaluated based on geological, geophysical, and geochemical research.

Furthermore, the Subsoil Resources Law provides that a license holder may request an extension of an existing license where extractable reserves remain upon the expiration of the primary term of the license, provided that the license holder is in material compliance with the license.

However, in accordance with the Law on Subsoil, mineral reserves in Russia are subject to mandatory state examination, and subsoil users cannot be granted a production license with respect to a field that was not examined. The state examination of reserves is conducted by subsidiary organizations of the Federal Agency on Subsoil Use, including the State Reserve Commission and Central Reserve Commission and its regional departments. If the commercial feasibility of certain reserves is approved by any such organization, the reserves are entered in the State Balance of Mineral Products. Once a subsoil user is granted an exploration, development or production license, it is required to file annual statistical reports reflecting changes in reserves. In addition, subsoil users' reserve reports are submitted annually for examination and approval by the Central Reserve Commission or its regional organizations or, if there has been a substantial change in reserves, by the State Reserve Commission.

Estimations of reserves, as examined by the state expert organizations and reflected in subsoil users' annual statistical reports, are accumulated in the State Balance of Mineral Products.

Finally, reserves of the combined A, B, and C1 classes in the Russian Standards equate with proved, probable and much of the possible reserves (3P) under the traditional SPE Standards.

4.2.4 Miscellaneous Standards

The Society of Petroleum Engineers (SPE), World Petroleum Council (WPC), American Association of Petroleum Geologists (AAPG), and Society of Petroleum Evaluation Engineers (SPEE) adopted a more sophisticated system of evaluating petroleum accumulations in 2007. This system incorporates the 1997 definitions for reserves, but adds categories for contingent resources and prospective resources.

Contingent resources are those quantities of petroleum estimated, as of a given date, to be potentially recoverable from *known* accumulations, but the applied project(s) are not yet considered mature enough for commercial development due to one or more contingencies. Contingent resources may include, for example, projects for which there are currently no viable markets, or where commercial recovery is dependent on technology under development, or where evaluation of the accumulation is insufficient to clearly assess commerciality.

Prospective resources are those quantities of petroleum estimated, as of a given date, to be potentially recoverable from undiscovered accumulations by application of future development projects. Prospective resources have both an associated chance of discovery and a chance of development.

The differences between the data obtained from these various estimates can be considerable, but it must be remembered that any data about the reserves of petroleum (and, for that matter, about any other fuel or mineral resource) will always be open to questions about the degree of certainty. Thus, in spite of the use of self-righteous word-smithing, proven reserves may be a very small part of the total hypothetical and/or speculative amounts of a resource.

The United States Geological Survey uses the terms technically and economically recoverable resources when making its petroleum resource assessments. Technically recoverable resources represent that proportion of assessed in-place petroleum that may be recoverable using current recovery technology, without regard to cost. Economically recoverable resources are technically recoverable petroleum for which the costs of discovery, development, production, and transport — including a return to capital — can be recovered at a given market price.

Unconventional resources exist in petroleum accumulations that are pervasive throughout a large area. Examples include tar sand bitumen (sometimes referred to incorrectly as extra heavy oil) and oil shale deposits. Unlike Conventional resources in which the petroleum is recovered through wellbores and typically requires minimal processing prior to sale, unconventional resources require specialized extraction technology to produce. Moreover, the extracted petroleum may require significant processing prior to sale (e.g. bitumen upgraders) (Speight, 2007, 2008). The total amount of unconventional oil resources in the world considerably exceeds the amount of conventional oil reserves, but is much more difficult and expensive to develop.

4.3 Benchmark Crude Oils

The price of petroleum means the spot price of either West Texas Intermediate crude oil as traded on the New York Mercantile Exchange (NYMEX) for delivery in Cushing, Oklahoma, or of Brent crude as traded on the Intercontinental Exchange (ICE, into which the International Petroleum Exchange has been incorporated) for delivery at Sullom Voe. The price of a barrel of oil is highly dependent on both its grade, determined by factors such as its specific gravity or API gravity and sulfur content, as well as location. The vast majority of oil is not traded on an exchange, but instead on an over-the-counter basis, typically with reference to a marker crude oil grade that is quoted via pricing. Other important benchmark crude oils include Dubai, Tapis, and the OPEC basket.

For the purposes of pricing, crude oil is generally classified based on the API gravity and sulfur content. For example, light crude oil has low density, low viscosity, and low sulfur content, making it easier to transport and refine and, therefore, more expensive to purchase. Sweet crude oil has sulfur content less than 0.5% by weight and is usually light crude oil, making it more expensive, but much easier to refine in a way that would meet environmental standards in developed countries.

A light crude oil is generally one with an API gravity of less than about 40. Brent crude oil has an API gravity of 38 to 39. Sweet crude is preferable to sour crude oil because, like light crude oil, it is also more suited to the production of the most valuable refined products. On the other hand, heavy crude oil has high density, high viscosity, and high sulfur content, making it more difficult to transport and refine and cheaper to purchase. Sour crude oil has a sulfur content above 0.5% by weight and is usually heavy crude oil, making it cheaper to purchase but more expensive to refine.

Heavy crude oil will typically have an API gravity of 20 or less — the higher the API gravity, the lower the density. Heavy crude oil is harder to handle because it is too thick to pump easily through pipelines unless diluted with light crude, and it is more expensive to refine to produce the most valuable petroleum products such as petrol, diesel, and aviation fuel.

Almost every oil field produces crude with a unique mixture of characteristics and it is, therefore, easiest to follow the prices of key benchmark varieties. The petroleum industry generally classifies crude oil by the geographic location it is produced in (e.g. West Texas Intermediate crude oil, Brent crude oil, or Omani crude oil), its API gravity (an oil industry measure of density), and by its sulfur content. In addition, the geographic location is important because it affects transportation costs to the refinery. *Light crude* oil is more desirable than heavy oil because it produces a higher yield of gasoline, while sweet oil commands a higher price than sour oil because it has fewer environmental problems and requires less refining to meet sulfur standards imposed on fuels in consuming countries. Each crude oil has unique molecular characteristics that are understood by the use of crude oil assay analysis in petroleum laboratories. Crude oil from an area in which the crude oil's molecular characteristics have been determined and the oil has been classified are used as pricing references throughout the world.

Because there are so many different varieties and grades of crude oil, buyers and sellers have found it easier to refer to a limited number of reference, or benchmark, crude oils. Other varieties are then priced at a discount or premium, according to their quality. Thus, crude oil is priced in terms of regional blends, each with different characteristics. Of these characteristics, traders follow certain blends, as they most reflect the overall value of oil, and therefore affect the way different blends are priced. These are essentially like a Consumer Price Index for different types of oil. Approximately 160 different types of crude that are traded around the world; the four primary benchmarks, of which these are priced internationally are:

1. Brent blend crude oil
2. West Texas intermediate (WTI) crude oil
3. Dubai crude oil
4. The OPEC Basket crude oil.

By way of clarification, a benchmark is, in the current context, a standard against which the properties of crude oil can be measured or compared. When evaluating the price of any crude oil it is important to compare the crude oil against an appropriate benchmark crude oil.

The Brent crude oil blend is based on the prices of Brent crude, which is a light, sweet crude oil and is actually a combination of crude oil from 15 different oil fields in the Brent and Ninian systems located in the North Sea. The API gravity is 38.3°, making it a light crude oil, but not quite as light as West Texas Intermediate crude oil, while it contains about 0.37% by weight sulfur, which makes it a sweet crude oil, but slightly less sweet than West Texas Intermediate crude oil. The Brent blend is ideal for making gasoline and middle distillates, both of which are consumed in large quantities in Northwest Europe, where Brent blend crude oil is typically refined. However, if the arbitrage between Brent and other crude oils, including WTI, is favorable for export, Brent has been known to be refined in the United States (typically the East Coast or the Gulf Coast) or the Mediterranean region. Brent blend, like West Texas Intermediate crude oil, production is also on the decline, but it remains the major benchmark for other crude oils in Europe or Africa. For example, prices for other crude oils in these two continents are often priced as a differential to Brent (i.e. Brent minus $0.50). Brent blend is generally priced at about a $4 per-barrel premium to the OPEC Basket price or about a $1 to $2 per-barrel discount to West Texas Intermediate crude oil, although on a daily basis the pricing relationships can vary greatly.

The West Texas intermediate (WTI) crude oil is the benchmark for oil prices in the United States based on light, low sulfur (0.24% by weight) West Texas intermediate crude oil, which remains the benchmark for oil prices in the United States despite the fact that production of this crude oil has been decreasing over the past two decades.

West Texas Intermediate crude oil is of very high quality and is excellent for refining a larger portion of gasoline. Its API gravity is 39.6 degrees, making it a light crude oil, and it contains only about 0.24 percent of sulfur, making a sweet crude oil. This combination of characteristics, combined with its location, makes it an ideal crude oil to be refined in the United States — the largest gasoline consuming country in the world. Most West Texas Intermediate crude oil is refined in the Midwest region of the United States, with some more refined within the Gulf Coast region. Although the production of WTI crude oil is on the decline, it still is the major benchmark of crude oil in the Americas. West Texas Intermediate crude oil generally priced at about a $5 to $6 per-barrel premium to the OPEC Basket price and about $1 to $2 per-barrel premium to Brent, although on a daily basis the pricing relationships between these crude oils can also vary greatly.

Dubai crude oil is a benchmark for Persian Gulf crudes, and is light yet sour crude oil. The OPEC crude oil Basket blend is OPEC's benchmark and is a weighted average of oil prices collected from various oil producing countries. This average is determined according to the production and exports of each country and is used as a reference point by OPEC to monitor worldwide oil market conditions. As of June 15, 2005 the basket was changed to represent the oil produced by OPEC members and is and is made up of 13 different regional oils, namely: Algeria's Saharan Blend, Angola's Girassol, Ecuador's Oriente, Indonesia's Minas, Iran's Iran Heavy, Iraq's Basra Light, Kuwait's Kuwait Export, Libya's Es Sider, Nigeria's Bonny Light, Qatar's Qatar Marine, Saudi Arabia's Arab Light, the United Arab Emirates' Murban, and Venezuela's BCF 17 crude oil.

As mentioned above, because WTI crude oil is a very light, sweet crude, it is generally more expensive than the OPEC Basket blend. Brent is also lighter, sweeter, and more expensive than the OPEC basket, although less so than West Texas Intermediate crude oil.

Since the marker crude system was introduced in the mid-1980s, there has been general industry acceptance that spot trade in these barrels acts as a barometer of the overall market level. Different grades of oil are priced on negotiable differentials to the marker grade. The rationale is that, in any market, the spot price represents the balancing point of supply and demand. Even though the volumes of oil that trade daily on a term contract basis between companies or governments are much bigger than those that trade on a spot basis, price is determined at the margin in the spot market.

For the last 20 years, price discovery in the oil market has been concentrated around three main regional crude oil benchmarks, also known as 'marker' crudes: West Texas Intermediate (WTI) from the United States, Brent Blend from the UK North Sea and Dubai, or Fateh, crude from the United Arab Emirates.

The relative value of different crude oils is determined by two main factors: location and quality. In the case of Brent crude oil and West Texas Intermediate crude oil, the crude produced is light and sweet. Both grades are produced in or near key oil consuming and refining centers. The twin advantage of quality and location mean that such a crude oil can command a relatively high price.

West Texas Intermediate crude oil is widely used in the pricing of US domestic crudes, as well as oil imports into the US. Brent crude oil has become the de facto international oil benchmark partly because of its location and partly because it is a good quality oil that can be used by a wide range of refineries. The physical value of North Sea Brent crude oil is widely used in benchmarking the bulk of oil from the North Sea, West and North Africa, Russia and Central Asia, as well as large volumes from the Middle East heading into western markets. Dubai, meanwhile, is a medium-to-heavy, low-sulfur crude oil that is typical of the grades produced in the Persian Gulf, but distant from consuming centers. As a result, it tends to sell at a lower price than Brent crude oil and West Texas Intermediate crude oil.

In recent years, the production level of each of the marker crudes has fallen; and in the case of Dubai, it has fallen drastically. Meanwhile, the proportions of heavier and sourer crudes that change hands in the term market have grown relative to light, sweet production (Swain, 1991, 1993, 1998, 2000). In fact, more than half the world's produced oil is heavy and sour in quality and this proportion is expected to increase (Chapter 2).

Middle Eastern crudes, typified by the heavy sour volumes that flow through the Strait of Hormuz, have grown in importance as a supply source for markets in both east and west. The dominant role of Saudi Arabia, the world's largest oil exporter, has been a factor in this, as the kingdom has the only immediately available spare capacity among OPEC member nations. Meanwhile, Russia has emerged as a major supplier to western markets after a period of decline in the early 1990s, but most of the oil exported from Russia is also relatively heavy and sour in quality.

In fact, the Russian benchmark crude, known as Russian Export Blend, is a mixture of several crude grades used domestically or sent for export. The Export Blend is a medium, sour crude oil with an API gravity of approximately 32 and a sulfur content of approximately 1.2% by weight.

While supply has typically become poorer in quality (Swain, 1991, 1993, 1998, 2000), with more heavy and high-sulfur crude oil in the mix, the demand for products has veered the other way; demand for light products has grown most rapidly. On the other hand, the demand for residual fuels from utilities has declined mainly because of substitution by gas. Meanwhile, the quality requirements for light products have become ever stricter. Sulfur levels in transportation fuels have tightened relentlessly — a trend that has accelerated since 2000 when leaded fuel was outlawed within the nations of the European Union. Furthermore, the sulfur content of gasoline and diesel are now below 50 ppm in Europe and the United States. Similar rules are being applied in many of the industrialized Asian countries and much of the developing world is heading in the same direction, although at a much slower rate.

This disparity between crude oil and product quality will continue to widen, although most petroleum economists concur that production from the North Sea and within the United States is on the decline relative to demand in the coming years considering that both Europe and North America more dependent on imported oil. These imports are likely to be increasingly supplied from the regions with the biggest reserves, such as the nations concentrated in the Persian Gulf, and are mainly sour in quality. Other countries with large reserves, such as Venezuela and Russia, also typically produce heavier sourer crudes.

As the world becomes critically more reliant on heavier and higher sulfur streams, it is the sour crude that is gradually assuming more importance in the price discovery process. Another problem is the declining production of benchmark crude oils. Thus, Urals crude in Europe and Mars crude oil blend in the United States have emerged as alternative benchmarks to their sweet counterparts Brent and West Texas Intermediate crude oil. This change in emphasis has been encouraged by the extreme volatility seen in recent months between light sweet and heavy sour crude oils, which can be extremely problematic for setting monthly official selling prices.

In summary, a benchmark crude oil (or a marker crude oil) is a crude oil or crude blend that is freely traded on a selected cash and futures market according to specified rules. West Texas Intermediate is traded only in the New York Mercantile Exchange in New York, and Brent crude oil is traded only in the International Petroleum Exchange; while Dubai-Oman crude oil is traded in Singapore.

However, much of the world crude oil trade does not involve benchmark crude oils. The purpose of trading in the benchmark is almost purely for monetary protection by energy industry users of other crude oils against future price movements that would harm their profit positions. In this sense, purchasing benchmark crude is a form of protection on owned or promised inventory of crude oil.

But while the two largest benchmarks, Brent and West Texas Intermediate, are light and sweet, most of the world's crude trade is in heavy sour crude oil. At best, the existing benchmarks provide some contra-price movement protection, but the Dubai-Oman crude oil blend was established as a Middle East benchmark to better estimate the market price for heavy, sour crude oil.

4.4 References

Aluko, M.E. Shell, Nigeria and Oil & Gas Reserves Revision - Sloppiness or Fraud? http://www.nigerdeltacongress.com/sarticles/shell_nigeria_and_oil__gas_reser.htm

ASTM D97. 2008. Standard Test Method for Pour Point of Petroleum Products. Annual Book of Standards. American Society for Testing and Materials. Philadelphia, Pennsylvania. Volume 05.03.

ASTM D4175. 2008. Standard Terminology Relating to Petroleum, Petroleum Products, and Lubricants. Annual Book of Standards. American Society for Testing and Materials. Philadelphia, Pennsylvania. Volume 05.03.

BP. 2008. BP Statistical Review of World Energy. www. bp.com/statisticalreview.

Campbell, C.J. and Laherrère, J.H. 1998. The End of Cheap Oil. Scientific American. 278: 78–83.

Speight, J.G. 2007. The Chemistry and Technology of Petroleum, 4th Edition. CRC Press, Taylor & Francis Group, Boca Raton, Florida.

Swain, E.J. 1991. Oil & Gas Journal. 89(36): 59.
Swain, E. J. 1993. Oil & Gas Journal. 91(9): 62.
Swain, E.J. 1998. Oil & Gas Journal. 96(40): 43.
Swain, E.J. 2000. Oil & Gas Journal. March 13.
US Congress. 1976. Public Law FEA-76-4. United States Congress, Washington, DC.

5

The Petroleum Culture

The definition of crude oil is confusing and variable (Chapter 1) and has been made even confusing by the introduction of other terms that add little, if anything to petroleum definitions and terminology (Speight, 2007; Zittel and Schindler, 2007). For example, lately it has been pointed out that there are different classification schemes: based on economic and/or geological criteria (Speight, 2007). The economic definition of conventional oil is "conventional oil is oil which can be produced with current technology under present economic conditions." The problem with this definition is that it is not very precise and it changes whenever the economic or technological aspects of oil recovery change. In addition, there are other classifications based on API gravity such as "conventional oil is crude oil having a viscosity above 17°API". However, these definitions do not change the definition stated elsewhere (Chapter 1) that is used through this book.

Whatever the definition, petroleum is a finite resource and for the past five decades, since the first postulations of resource depletion (Hubbert, 1962), many scientists and engineers have warned of its continued depletion leading to exhaustion of the available resources.

It is assumed with some degree of justification that the extraction of oil, like that of other nonrenewable resources, will follow a parabolic (bell-shaped) curve over time and the only unknown is the characteristics of the slope on the depletion side of the curve.

Crude oil is no different than many other natural resources insofar as production rises quickly at first and then gradually slows until approximately half the original supply has been exhausted; at that point, a peak in sustainable output is attained and production begins an irreversible decline until it becomes too expensive to recover the remaining in-ground material. In accordance with Hubbert's postulate, many scientists and engineers believe that the midway point in the depletion of the original world petroleum inheritance has been reached and severe depletion of the resource is already underway. On the other hand, there are opposing claims to the effect that the peak of oil production is still a year or maybe decades away, and the world has not yet reached the depletion side of the bell curve.

However, it is certain that the reserves of crude oil are being depleted and the discovery of new fields has slowed. In fact, many scientists and engineers conjecture that new giant oil reservoirs are waiting to be discovered while the opposing view that there are no new giant oil reservoirs holds at least equal weight. If this latter opinion is the case, as the world sinks into crude oil oblivion, it is worth considering the phenomenon of the petroleum culture since this has a high influence on petroleum economics.

In part, many ancient cultures collapsed because of their inability to maintain important resources that allowed society to flourish (Tainter, 1988). Indeed, the development of the modern world focused and became heavily dependent on the recovery and use of petroleum. As of late, this has taken a serious turnabout because some of the most promising new oil discoveries have turned out to be very disappointing in terms of the amount of the resources contained in these fields (Cooper and Pope, 1998).

The economics of a petroleum-based culture involves energy resources and energy commodities, including:

1. Forces motivating firms and consumers to supply, convert, transport, use energy resources, and to dispose of residuals
2. Market structure and regulatory structures
3. Distributional and environmental consequences; economically efficient use.

Such a culture should recognize: (1) energy is neither created nor destroyed but can be converted among forms, and (2) energy comes from the physical environment and ultimately returns there. In addition, energy demand in such a culture is derived from preferences for petroleum products that depend on properties of conversion technologies, as well as cost of the technologies. Petroleum resources are depletable or renewable and storable or non-storable. Crude oil is a depletable resource and market forces may guide a transition back to renewable resources.

As of year-end 2008, global crude oil supply and demand was in tight balance, resulting in the rapid increases in crude oil prices and price volatility between 2007 and 2008. As the benchmark West Texas Intermediate crude oil exceeded $145/bbl in mid-2008, the world began to see a moderation of demand growth caused by high prices. Crude oil demand actually decreased in the United States, Western Europe, and Southeast Asia, while demand growth moderated in the fast economic growth countries such as Brazil, Russia, India and China. Global crude oil demand growth of 2% per year, which had characterized 2000–2006, was cut nearly in half a couple years larer by the cumulative impacts of higher petroleum prices, lower global GDP, and mandated production and use of biofuels. Claims that high oil prices were being driven by futures market speculation (dry barrel participants), profiteering by oil companies and the withholding of supply by OPEC are conspicuously absent from this government assessment. During 2007 to 2008, geopolitical concerns over some major oil supplier output (Nigeria, Iraq, and Venezuela) contributed to oil price volatility.

Over the longer term, primary global energy demand is expected to increase by 50–60% by the year 2030, driven primarily by population growth and the desire for better living standards. Conventional oil and gas alone are unlikely to satisfy the demand growth in its seminal 2007 report, and called on the development of supplemental energy sources such as clean coal, biofuels, alternative renewal energy, nuclear, and non-conventional fossil fuels (Chapter 9).

Thus, in early in 2009, the state of oil-supply predictions covers a wide range of opinions and it is important to assess the main alternatives (Speight, 2007, 2008).

However, an even greater problem may be that an increasing number of politicians and company decision-makers sense that technology will resolve this issue as it has done before and will do so again. This line of thinking may be courting disaster.

5.1 The Petroleum Culture

The production of energy from various sources has played a major role in the history of human culture (Henry, 1873; Forbes, 1958a, 1958b, 1959, 1964; Hoiberg, 1964). These developments have increased the comfort, longevity, and affluence of humans. It was in the 20th century that petroleum became the predominant hydrocarbon energy source and has played major role in the advancement of human culture ever since (Munasinghe, 2002); there continues to be a strong connection between energy and economic activity for most industrialized and developing economies (Sadorsky, 1999; Hall *et al.*, 2001; Tharakan *et al.*, 2001; Smulders and de Nooij, 2003; Speight and Cockburn, 2008).

5.2 Oil in Perspective

5.2.1 History

The use of petroleum and petroleum derivatives had been known for millennia (Cobb and Goldwhite, 1995). However, the petroleum culture (or Petroleum Age by analogy to the Stone Age, Bronze Age, and Iron Age) started in Titusville, Pennsylvania on August 27, 1859 when self-styled "Colonel" Edwin L. Drake, the first wildcatter, struck oil and transformed the northwestern portion of the state into an economic boomtown.

Briefly and by way of explanation, a wildcatter is a person who drills searches for oil and drills wells in areas that are not known to be oil fields. The term originates from drillers in west Texas who would clear prospective fields of wildlife, including the ever-present feral cats. Generally, the drillers would hang the pelts of the dead cats in the rigging of the finished derricks, from which these exploratory wells (wildcat wells) and later those who prospected for them, eventually took their name.

Colonel Drake was prompted by the efforts of George Bissell, a Wall Street attorney and investment broker, to extract rock oil (petroleum) from the ground to serve as an illuminant to replace whale oil, which was becoming more and more difficult tom obtain (Henry, 1873). Benjamin Silliman developed the process of refining the oil to a usable flammable liquid, and the final illuminating oil product became known as kerosene.

In 1865, John D. Rockefeller exploited the discovery or petroleum and founded Standard Oil based on a fledgling refinery business in Cleveland, Ohio. Pipelines were built to bypass the high cost of delivering the wooden barrels (formerly whiskey casks) and eventually these pipelines linked the Pennsylvania fields to railroads which sent the oil and kerosene to markets in the industrialized northeastern United States. During this time, Rockefeller bought and consolidated the supply and distribution sectors under one company banner making Standard Oil an effective and self-sustaining company.

In 1873, Russia also commenced development of a modern petroleum industry in Baku (Caucasus region). Oil has been produced from this region for several hundred years — perhaps even millennia — but there was not regular industry in 1873. At the time, the Nobel family, of Nobel Prize fame, purchased an embryonic refinery and commenced to increase oil production as well as distribution with the first efficient tanker ship.

A few years later, Marcus Samuel, the son of a London shell merchant, established an extensive trade business with Far Eastern connections to ship kerosene to the rest of the world. He joined forces with the Rothschild family, and in 1891, he was shipping the illuminant oil from the Caucasus through the Suez Canal to Hong Kong and the Dutch East Indies. At the same time, Royal Dutch Oil was exploiting far eastern oil, and eventually Samuel's Shell Oil and the Rothschild family joined the company to become known as Royal Dutch Shell.

The production of automobiles by Henry Ford in 1896 spurred demand for a liquid fuel to power the vehicles, and, to meet that demand, prospecting for oil led to the discovery of the Spindletop field in Beaumont, Texas, in 1900 followed by discoveries of oil fields in California and in Oklahoma. This facilitated the birth of more oil companies with names including Sun, Texaco, and Gulf. Unfortunately, as had happened in Pennsylvania, over-production in Texas resulting from poor reservoir management drove prices down very quickly and the profit interests in petroleum were markedly diminished.

However, this did not kill the need for crude oil. Although it did suffer setbacks in the early development of various fields, the industry was resilient and expanded phenomenally as the demand for petroleum products grew. The need for kerosene as a fuel for illuminating lamps varied as electricity and the incandescent lights grew to maturity and became readily available (Davis, 2003). The deciding factor often cited for the expansion of the petroleum industry is the development of the automobile. However, this development, plus the need for fuel for military vehicles (including aircrafts), assured the future of the petroleum industry. As a result, the powerhouse of petroleum reserves was starting to emerge.

5.2.2 The Middle East Emerges

Several countries of the region defined as the Middle East are classed as oil-rich because of the reserves of oil contained in formations below the surface the earth, but within their receptive borders. However, to put the current circumstances in the Middle East into perspective, it is imperative to observe the history of oil discovery and production both in the domestic United States and in other countries.

Various exploratory ventures were launched in Persia (later Iran) at the turn of the 20th century. The first site chosen was at Chiah Surkh amidst the mountains of what is now northeastern Iraq. Oil was discovered in 1903 after a year of drilling and constantly challenging supply and political problems leading to host of financial problems. The British government got into the act through the intermediary of Thomas Boverton Redwood, a member of the Admiralty's Fuel Oil Committee, who brokered a bailout under the auspices of Burmah Oil, a British source of crude oil in Rangoon.

Since the advent of fuel oil, the British navy recognized that the future of ships lay in oil-powered vessels, and, because the Persian wells were proving reliable, the role of security would become the responsibility of the British navy. The formal nature of the partnership was defined by Lord Lansdowne in the House of Lords in 1903, with the Lansdowne Declaration:

> The British government would regard the establishment of a naval base or a fortified port in the Persian Gulf by any other power as a very grave menace to British interests, and we should certainly resist it with all the means at our disposal (Yergin, 1991).

The name of the concession was the Concession Syndicate (Wilson, 1941), but it would grow to become British Petroleum (Yergin, 1991).

In 1906, the Concession Syndicate capped the northern fields and moved the operation south where access was easier and prospects better. When the Shah was deposed and a new or Parliament took over, the British and the Russians decided to section Persia. As a result of this sectioning, the Russians would cover the north and the British would cover the south. Oil started to flow in 1912 by means of a 138-mile pipeline to a refinery in Abadan on the Shatt-al-Arab waterway between what would eventually become the border of Iran and Iraq.

In order to deny the ever-increasing German navy the access to this oil, the British government would buy 51% of the Abadan refinery concern and, in addition, the British fleet was granted a 20-year contract for fuel oil. The first quasi-nationalized company takeover had occurred.

World War I broke out in 1914, and from the beginning, it was a war fought with land machines and aircrafts powered by petroleum products. This served to increase the demand for crude oil and carried the war to areas other than the Western Front. The demand for oil was such that countries that had no interest in the war became embroiled in the conflict as the right to oil were fought over with untold violence and, not forgetting the obvious, political maneuvering (Yergin, 1991).

In the United States, the demand for automobiles was underway (Setright, 2004) and the effect on gasoline demand was appreciable leading to concern about depletion of the domestic supply of petroleum. As a result, a geological expedition was dispatched to Iraq to determine if in fact there was oil. Drilling began in Baba Gurgar near Kirkuk, and, within six months the drillers struck oil.

At this time, oil was seen as a truly rich resource, and between World War I and World War II, thoughts of nationalization spread. In fact, the Shah of Persia seized the assets of the Anglo-Persian oil Company in November 1932, leading him to seek a greater percentage of wealth from oil. This led to a new agreement in which Persia was guaranteed a fixed royalty, regardless of market price of oil.

The 1930s depression had affected the Saud dynasty as it had most of the world and as such the king of Saudi Arabia, founded on the unification of various tribal territories by Muhammad Ibn Saud aided by followers of Muhammad Ibn Abdul Wahab, also saw the potential wealth that lay being an oil industry. During a search for a viable water source, several promising oil sites were noted, and, after oil was discovered in Bahrain, Ibn Saud decided it would be in his country's best interest to develop any oil fields that could be identified.

In the meantime, the Amir of Kuwait, Sheikh Ahmad, witnessed the events in Bahrain and Saudi Arabia and also decided that oil was to be a large part of his country's future. In quick succession, oil was struck in Kuwait on February 28, 1938, followed in March by discoveries in Saudi Arabia, and, in April 1939, King Ibn Saud opened the pipeline feeding the terminal at Ras Tanura and oil flowed to a waiting Socal tanker.

Unfortunately, the best-laid plans were not executed and World War II caused the cessation of virtually all crude oil production in the Persian Gulf.

In 1943, President Roosevelt authorized Lend-Lease assistance to Saudi Arabia to tie US national security to Middle Eastern oil. Further measures were taken to acquire ownership of foreign reserves through the newly created government entity known as the Petroleum Reserves Corporation, but the oil companies reacted strongly to the idea of nationalization of American corporations and the plan was abandoned.

Before 1956, Egypt did not possess oil wealth, but it did have the Suez Canal through which flowed two-thirds of Europe's oil supply. In 1954, King Farouk was deposed and General Gamal Abdel Nasser rose to power took on the role of president. The general was faced with national economic hardship and decided to raise revenues from transit fees through the canal. This required that he had to get control of the canal by displacing the Suez Canal Company, which was not an Egyptian venture but joint venture of Britain and French.

General Nasser often advocated the use of oil as an economic weapon but failed to garner support from many of the oil-producing Arab countries. After General Nasser died in 1970, Anwar Sadat ascended to the position President, and, in 1972, he had tried to induce the oil-producing Arabian nations to use the oil weapon, but King Faisal of Saudi Arabia disagreed. His rationale was that Egypt had been politically unstable for some years, and while Egypt was moving closer to trade and weapons agreements with Russia, the United States (linked to Saudi Arabia by similar agreements) was projected not to need Arab oil until 1985; therefore, the oil weapon would only hurt Saudi Arabia.

As it turned out, by 1973 (more than a decade earlier that projected by King Faisal), they had become dependent on Middle Eastern oil by 1973, which placed Saudi Arabia firmly in a position of control and instituted an embargo because of the support of Israel by the United States. As a result, the oil weapon fell into place and was used with great effect. Consumers in the United States were finally confronted with the real nature of their energy dependence, which bolstered the emerging environmental movement. And, finally, the oil-producing countries had tasted power and were ready to consolidate their position. The impact on the United States was that inflation became embedded in the economy, based in large measure by the increasing price of crude oil.

The oil-producing countries strengthened their position but had, at this time, not achieved full cartel status. In fact, soon after this, all of the oil-producing countries were at maximum capacity, with the exception of Saudi Arabia.

Furthermore, in 1974, the Saudi Arabia took 60% of Aramco (Exxon, Mobil, Texaco, and Chevron), and finally, in 1976, Aramco was disbanded and the Saudi government (i.e. the Saud family) took control of the estimated 149 billion barrels of reserves. The new arrangement allowed the companies to have access to 80% of production while being compensated 21 cents a barrel for operating the production services.

The post-embargo period was represented best by the scramble for oil everywhere by everyone at any price. While that was happening, exploration and development was targeted at western countries where nationalization was not a concern.

In 1969, Phillips Petroleum struck oil in the North Sea with British Petroleum (now BP) also striking oil in 1970; Shell and Exxon discovered the Brent field joined them in 1971. By 1975, oil flowed from offshore pipelines to British refineries. However, the British government considered nationalizing their portion of the field and formed the British National Oil Corporation (BNOC), which held title to the government's concession and the right to buy 51% of North Sea production.

By the 1980s oil was considered to be a commodity and became a tangible asset that was traded like any other commodity, such as gold and money. As domestic production increased in the United States, Exxon terminated the Colony Oil Shale project and also cut back on oil exploration in 1982. In addition, shipping became more efficient and the Alaska pipeline boosted output that contributed to an excess of oil due to the increased supply from non-OPEC countries, which even surpassed oil production from the OPEC members.

By this time, the OPEC member countries had changed the way that oil was handled on the market and had finally become a cartel insofar as they price and production. In short, the market economy of oil had made it a true commodity. The market shifted from OPEC to the New York Mercantile Exchange in 1983, and the futures market now included oil as a speculative commodity. With this shift, oil companies moved from production to speculation on the open market, but the pivotal moment was when the benchmark crude went from Arab Light crude oil to West Texas Intermediate (WTI) crude oil, reversing the move made two decades earlier when Arab Light crude oil had replaced Texas Gulf Coast crude oil as the benchmark crude oil.

In 1984, the OPEC member countries found that the quota system they had instituted was flawed. Most member countries had programs that, if they were not an outright sale of oil, then it was through barter trade for military hardware and industrial goods and the crude oil market was in a state of oversupply. The OPEC members had the option to lower prices or continue to prop up the prices with a continued loss of market share.

However, at a meeting in fall of 1985, the members collectively declared a price war. After the author's return from Iraq, where he had a meeting with Vice President Sadam Hussein, the author warned of such an event but the idea was dismissed and even called ludicrous by a so-called energy economist for an academic institution. Yet, it happened. Oil was a truly an economic weapon and the result was economic warfare.

The end of the 1980s conveyed the state of affairs in the oil industry. The member-countries (i.e. the producers) had taken over crude oil production and the oil companies were reduced to the rank of contractors. Even though consumers fought back and took their purchasing power to producers where the price was right, stability had not been reached.

5.2.3 Recent History

Between 1991 and 2003, there were notable events that shaped the oil industry and international economies. The spike in prices after the Gulf War reached a high, but quickly shrunk after the United States released some of the Strategic Petroleum Reserves. The former Soviet Union broke up and the former republic's oil-producing countries started to court the world market, which further reduced oil prices. However, the OPEC members raised prices while raising production to the highest level in decades. Kuwait defied the OPEC quotas to finance reconstruction and prices immediately decreased.

Prices peaked during the 1995 to 2000 period, and supply began to swell when Iraq re-entered the market under the UN-sanctioned Food-for-Oil program, which drove prices down. OPEC adjusted by trying to match cheating with increased production, and oil prices plummeted to pre-embargo levels due to anticipated Asian consumption, which never materialized because of the economic crisis there. Between 1999 and 2000, cuts by the OPEC members and increased demand, due to cold weather and resupply of low stocks, increased the price of oil.

Recession in the US contributed to price declines in 2000, and the September 11th attacks of 2001 pushed them even further down. In 2002, with Venezuela in turmoil, prices went even lower, causing companies to immediately stockpile for seasonal and financial reasons. By winter of 2003, prices had rebounded due to cold weather and renewed talk of hostilities with Iraq. Profits for companies were at an all-time high as stocks were sold in preparation for the coming war with Iraq.

The reintroduction of Iraq as an oil producer may have profound effects in global oil prices, but the prospects for Iraqi face political impediments (Shafiq, 2009a, 2009b). There are no giant oil fields being discovered and all of the previous economic issues have forced the petroleum industry to streamline the distribution process to soften the blow of supply interruptions. However, expansion of global demand for crude oil is expected to increase from 80 to 85 million barrels per day to approximately 120 million barrels per day by 2020. Therefore, cooperation is vital to avoid disruption and the ensuing economic hardship. Furthermore, the world will not be replacing crude oil as the primary energy source any time soon, no matter how much energy is projected to come from alternate sources.

In summary, this invokes the concept of geopolitics or the politics of oil: a friend today could be an enemy tomorrow.

5.3 The Seven Sisters

In the past, there have been frequent references to a group of oil companies generally referred to as the seven sisters, from a phrase first made popular by Italian oil tycoon Enrico Mattei: Exxon (Esso), Shell, BP, Gulf, Texaco, Mobil, and Chevron (Socal) (Sampson, 1975).

The rise of the seven sisters can be attributed to happenstance starting in 1904 when Calouste Gulbenkian, an Armenian businessman, reported to the Turkish Sultan Abdul Hamid on the oil potential of the then Turkish provinces. The Sultan, realizing that oil was a worthy possession, transferred large areas into his own personal possession. In 1914, the Iraq Petroleum Company was formed as a result of an agreement between British and Dutch companies. In 1928, 95% of the Iraq Petroleum Company was divided equally between the British (BP), the Dutch (Shell), the French (CFP, the Compagnie Française Pétrole), and a Rockefeller-controlled American group (Exxon and Mobil); the remaining 5% went to Mr. Gulbenkian, and the other 95%.

In 1908, the Anglo-Persian Oil Company, which eventually evolved into BP, discovered oil in Iran, and shortly thereafter in 1911, Winston Churchill, First Lord of the Admiralty, used Government money to buy half of the company on behalf of the Royal Navy. Churchill also decided that new British battleships would be fueled by oil rather than coal, and the Iranian supplies were very valuable to the British in World War I.

American companies, in the light of vast oilfields being discovered in Texas and California, were unwilling to explore abroad. However, the US government began to use considerable political and economic pressure to try to force American companies into the European-dominated consortia in the Middle East. In the 1920s, new fields came on line and there were serious concerns about excessive supplies of oil. By 1928, there were negotiations between BP, Shell, and Exxon leading to the Achnacarry Agreement, which set out working principles to avoid competition at the marketing end of the oil industry. The agreement had to exclude the US domestic market because of the new anti-trust legislation, but as a consequence of the Achnacarry Agreement, each large company could feel that it would be able to negotiate a market share for its oil without the danger of a price crash.

Briefly, the break-up of the Standard Oil in 1911 had effectively warned off any overt attempts at controlling the large American market, but the same constraints did not apply to the rest of the world.

After 1928, the era of the great Middle East oil strikes began, though Middle East production remained low. On June 1, 1932, SoCal (now Chevron) struck oil in Bahrain, the first strike in the Arabian Peninsula. In 1933, BP extended its Iranian lease for another 60 years and Gulf joined with BP to explore a Kuwaiti concession in 1934. In 1938, Gulf and BP discovered oil in what was to become the Burgan field in Kuwait, and Chevron struck oil in Saudi Arabia, which was marketed through Texaco's global sales network under the name Caltex, while the Saudi part of the partnership became known as *xi* (Arabian American Oil Company).

On October 3 1930, an independent wildcatter discovered the giant East Texas field, and more crude oil began to flow into the US domestic market in 1931. The oil companies began to buy up leases in the new fields, but the quantity of oil was too great to be absorbed easily, and the East Texas fields were soon producing a million barrels a day, one-third of all United States production.

The major oil companies had control over crude oil refining and product marketing through a system that was already tightly controlled. The companies could, in fact, state what they would pay for crude oil from the new fields and, as the new oil began to flow, that price was about $0.70 per barrel. At this time, the oil companies were concerned about overproduction because it was clear that pumping oil too rapidly from a field could damage long-term production. As a result, in November 1932, the Texas legislature passed the Market Demand Act, which defined prohibitable waste as any production that was in excess of market demand. As soon as the Act became law, the majors dropped their offering price to $0.25 per barrel on January 1933, and then to $0.10 per barrel. Production was cut dramatically as market demand dropped. The major companies then stepped in and bought up tens of millions of barrels at extremely low prices before many of the independents went out of business.

In 1935, the Congress of the United States passed an Interstate Compact to Conserve Oil and Gas, as well as the Connally Act, which assigned production quotas to each State and the legislation was used to hold down production in order to maintain stable prices. Another impact on United States domestic oil supplies was the policy decision to serve the domestic market largely from American oil wells so that the United States would not become too dependent on foreign oil and the cheap foreign oil was kept out of the United States while domestic fields were being depleted.

In 1947, oil profits were high. Saudi Arabian crude oil cost $0.19 a barrel plus $0.21 royalty, and Bahrain oil cost $0.10 a barrel plus $0.15 royalty. Consumers were paying $1.80 a barrel and more for that same crude oil. Profit margins increased in the 1950s and 1960s while production costs decreased. The volume of oil shipped increased, and the advent of supertankers decreased shipping costs while selling prices were increased. In the late 1960s, Middle East oil that was delivered to Europe and the United States at $2.00 or more per barrel had cost much less to produce and transport; some estimates indicate that the combined production and shipping costs were on the order of $0.40 per barrel.

By the end of the 1960s, the seven sisters remained the dominant companies in world oil. Between 1960 and 1966, their share of oil production outside North America and the Communist countries had actually gone up from 72 to 76%, leaving only 24% for all other companies (Penrose, 1969). The seven sisters had built themselves up into some of the biggest corporations in history, primarily through the ownership of concessions in developing countries, and predominantly in the Middle East. As a result, refinery complexes rose up along the coastlines and it even appeared that the companies inhabited a place between governments and business (Hartshorn, 1962).

By 1970, 94% of domestic oil reserves in the United States were held by only 20 companies. The top eight oil companies in terms of their holdings of domestic oil reserves were also the top eight in production, the top eight in refining, and the top eight in marketing: Exxon, Mobil, Texaco, Chevron, Gulf, Shell, Amoco, and ARCO.

In 1972, two of the seven sisters, Exxon and Shell, dominated the world's oil as well as the petroleum industry (Table 5.1). Mobil was long regarded as Exxon's little sister, dependent on her bigger rival both for advice and for oil. Now, Exxon owns Mobil, and the name may soon disappear from the ExxonMobil logo.

5.4 Reserve Estimates

Most estimates of the quantity of conventional oil resources remaining are based on the opinions of geologists and other scientists or engineers familiar with a particular region (Table 5.2) (Chapter 2).

The ultimate recoverable resource (URR) is the total quantity of oil that will ever be produced, including the nearly one trillion barrels extracted to date (Chapter 1). Recent estimates of ultimate recoverable resource for the world have tended to fall into two categories. Lower estimates indicate that the ultimate recoverable resource is no greater than about 2.3 trillion (2.3×10^9) barrels, and may even be less (Campbell and Laherrère, 1998). A higher estimate of the ultimate recoverable resource is 3 trillion (3.0×10^9) barrels and may be as much as 4 trillion (4.0×10^9) barrels of oil (USGS, 2000, 2003). About half of the approximate 1.4 trillion (1.4×10^9) barrels that are predicted by the United States Geological Survey to remain to be discovered are from new discoveries and about half are from

Table 5.1 The world's 12 largest manufacturing corporations ranked by assets and sales in 1972.

Rank	Company	Assets ($000)	Sales ($000)	Rank
1	Exxon*	21,558,257	20,309,753	2
2	Royal Dutch/Shell*	20,066,802	14,060,307	4
3	General Motors	18,273,382	30,435,231	1
4	Texaco*	12,032,174	8,692,991	10
5	Ford	11,634,000	20,194,400	3
6	IBM	10,792,402	9,532,593	7
7	Gulf*	9,324,000	6,243,000	12
8	Mobil*	9,216,713	9,166,332	8
9	Nippon Steel	8,622,916	5,364,332	17
10	ITT	8,617,897	8,556,826	11
11	BP*	8,161,413	5,711,555	15
12	Socal (now Chevron)*	8,084,193	5,829,487	14

*The Seven Sisters.
Source: *Fortune Magazine*, May and September 1973.

reserve growth. The latter describes the process by which technical improvements and correction of earlier conservative estimates increase the projected recovery from existing fields. This relatively new addition to the United States Geological Survey methodology is based on experience in the Unites States, as well as in other well-documented regions. The new totals essentially assume that petroleum reserves everywhere in the world will be developed with the same level of technology, economic incentives, and efficacy as in the US. Time will tell the extent to which these assumptions are realized.

5.4.1 Historical Variation of Reserve Estimates

Currently, the United States consumes roughly 20 million barrels per day with about 11.7 million barrels per day in imports.

Table 5.2 Estimates of the remaining reserves of crude oil (Hakes, 2000).

Source	Volume (Trillions of Barrels)
USGS, 2000 (high)	3.9
USGS, 2000 (mean)	3.0
USGS, 2000 (low)	2.25
Campbell, 1995	1.85
Masters, 1994	2.3
Campbell, 1992	1.7
Bookout, 1989	2.0
Masters, 1987	1.8
Martin, 1984	1.7
Nehring, 1982	2.9
Halbouty, 1981	2.25
Meyerhoff, 1979	2.2
Nehring, 1978	2.0
Nelson, 1977	2.0
Follinsbee, 1976	1.85
Adam and Kirby, 1975	2.0
Linden, 1973	2.9
Moody, 1972	1.9
Moody, 1970	1.85
Shell, 1968	1.85
Weeks, 1959	2.0
MacNaughton, 1953	1.0
Weeks, 1948	0.6
Pratt, 1942	0.6

OPEC producers account for 2/5 of this input and Persian Gulf states is 1/5 of that amount with Saudi Arabia having the largest block at 1.5 million barrels per day. Other large contributors are Canada (1.9 barrels per day), Mexico (1.5 million barrels per day), and Venezuela (1.4 million barrels per day). Iraqi oil could replace any of these importers, but it would only become likely if there was a major disruption like the kind that happened in Venezuela prior to Operation Iraqi Freedom and recently in Nigeria (.5 million barrels per day US imports). In both of those cases, OPEC took up the slack and continued providing the 2.5 million barrels per day of Iraqi production. Most of the shortfall was covered by Saudi Arabia. Realistically, Iraqi crude will find Asian and European markets with the US augmenting supply by courting it in negligible amounts.

While most of the world's attention is on Iraq and the Middle East, key events have occurred elsewhere that could affect markets in ways that reduce the impact of the return of Iraq to the market. Canada, which vies for the major supplier to the United States with Saudi Arabia at around 2 million barrels per day, increased its reserves from 4.9 billion barrels by 175 billion barrels of bitumen (for conversion to synthetic crude oil) from tar sand deposits. In terms of heavy oil, overall increases in efficiency and industry advances could add as much as 125 billion barrels to global reserves.

On the other hand, with proven reserves of 48 billion barrels, Russia has once more entered the fray. The Druzhba pipeline supplies oil to Europe and the aging terminal at Novorossisk on the Black Sea is still a consistent outlet. However, civil unrest has plagued the area since the fall of the Soviet Union and the Bosphorus Straits is not the easiest water for large carriers to negotiate and may become restricted should the Turkish government deem it. However, a pipeline bypass of the Bosphorus is underway to carry Caspian oil from Baku to Ceyhan in Turkey. In addition, Asia forms a ready market for oil from the Siberian fields and the possibility of a pipeline to China and Japan has been considered. In the meantime, China is seeking to develop its own heavy oil fields.

5.4.2 Patterns of Use

The best-known model of oil production was proposed by Marion King Hubbert, who proposed that the discovery, and production, of petroleum over time would follow a single-peaked, symmetric bell- shaped curve with a peak in production when 50% of the ultimate recoverable resource had been extracted (Oil Fields, 2008).

Hubbert saw that the facts of geology and the realities of physical pumping from underground meant that oil fields typically yield an increasing volume of oil, the volume then stops increasing (production hits a peak), then the volume that can be pumped out of the ground gradually decreases again. Put on a graph, production from an oil field makes a bell curve, starting low, slowly climbing, peaking, slowly descending, and finishing low. The high point on the curve is when half of the original reserves that are easily able to be pumped out of the ground in national or global oilfield resources are gone.

Knowing the shape of the production curve for oil fields meant Hubbert could take the figures on the known size of the oil reserves, how much oil had been pumped out of all the USA oil fields already, how much that still remained was recoverable, and fit these figures into a single oilfields of the United States production curve, representing the life and death of the totality of oil fields in USA. The results showed that oil production in the United States would increase in volume every year until 1970. It would then be at the peak of production, and decline thereafter.

In 1956, Hubbert predicted in that oil production in the United States would peak in 1970, which in fact it did (Hubbert, 1962). Hubbert also predicted that the US production of natural gas would peak in about 1980, which it did, although it has since shown signs of recovery. He also predicted that world oil production would peak in about 2000. There was a slight downturn in world production in 2000, but production in the first half of 2008 is running above the rate in 2000.

However, Hubbert's peak is for conventional crude oil and does not include oil produced for heavy oil fields, nor does it include any potential effects from liquids produced form non-conventional sources such as tar sand deposits, coal, oil shale, or liquids from natural gas.

In the past decade, a number of pseudo-sages have made predictions about the timing of peak global production using several variations of Hubbert's approach. Various forecasts of the year of the global peak have ranged from one predicted for 1989 (made in 1989) to many predicted for the first decade of the 21st century to one as late as 2030 (Campbell and Laherrère, 1998). Their predictions begin with an a priori assumption about the volume of ultimately recoverable oil. Most of these studies assumed world ultimate recoverable resource volumes of roughly 2 trillion barrels, and that oil production would peak when 50% of the ultimate resource had been extracted. In comparison, the United States Geological Survey low estimate (which they state has a 95% probability of being exceeded) is 2.3 trillion barrels. One analysis fitted the left-hand side of Hubbert-type curves to data on actual production while constraining the total quantity under the curve to two, three, and four trillion barrels for world ultimate recoverable resource. The resultant peaks were predicted to occur from 2004 to 2030.

Other forecasts for world oil production do not rely on such curve-fitting techniques to make future projections and/or a priori assumptions about ultimate recoverable resource. According to the most recent forecast by the US Energy Information Agency, EIA, world oil supply in 2025 will exceed the 2001 level by 53% (EIA, 2003). The EIA reviewed five other world oil models and found that all of them predict that production will increase in the next two decades to around 100 million barrels per day, substantially more than the 77 million barrels per day produced in 2001. Several of these models rely on the new United States Geological Survey estimates of ultimate recoverable resource for oil. It should be noted that almost all oil-supply forecasts for which we are able to examine the predictions against reality had a dismal track record, regardless of method. Most recent results of curve-fitting methods showed a consistent tendency to predict a peak within a few years, and then a decline, no matter when the predictions were made (Lynch, 2002). It is now a well-established fact that economic and institutional factors, as well as geology, were responsible for the US peak in production in 1970 (Kaufmann and Cleveland, 2001), forces that are explicitly excluded from the curve-fitting models. Thus, the ability (or the luck) of Hubbert's model and its variants to forecast production in the 48 lower states accurately cannot necessarily be extrapolated to other regions. It is too early to tell.

Economic forecasts fare no better in explaining US oil production in the lower 48 states. In the period after the Second World War, oil production often increased as oil prices decreased, and vice versa (Kaufmann, 1991), a behavior that is exactly the opposite of predictions of economic theory, which also assumes that oil prices will follow an *optimal path* towards the choke price (i.e. the price at which demand for oil falls to zero and the market signals a seamless transition to substitutes). In fact, even if such a path exists, prices may not increase smoothly because empirical evidence indicates that producers respond differently to price increases than they do to price decreases (Kaufmann and Cleveland, 2001). Significant deviation from basic economic theory undermines the de facto policy for managing the depletion of conventional oil supplies — a belief that the competitive market will generate a smooth transition from oil.

5.4.3 Energy and the Political Costs of Oil

The future of oil supplies is normally analyzed in economic terms. But, the economic terms are likely to be dependent on other costs. In earlier work, we summarized the energy costs of obtaining US oil and other energy resources and found, in general, that the energy returned on energy invested (EROI) tended to decline over time for all energy resources examined. This includes the energy cost of obtaining oil by trading energy-requiring goods and services for energy itself (Cleveland *et al.*, 1984). For example, the EROI of oil in the US has decreased from a value of at least 100 to 1 for oil discoveries in the 1930s, to about 17 to 1 today for oil and gas extraction. Similar types of estimates for other parts of the world are unknown, although heavy oil in Venezuela and tar sand bitumen in Alberta require a very large part of the energy produced as well as substantial supplies of hydrogen from natural gas to make the oil fluid. The very low economic cost of finding or producing new oil supplies in the Arabian Peninsula implies that it has a very high energy-returned-on-energy-invested value, which in turn supports the probability that productivity will be concentrated there in future decades. Alternative liquid fuels such as ethanol from corn have a very low energy-returned-on-energy-invested. n energy-returned-on-energy-invested value much greater than 1 to 1 is needed to run a society, because energy is also required to make the machines that use the energy, feed, house, train, and provide health care for necessary workers.

The means by which the various factors will play out over the next few decades is extremely important but also difficult to predict. Most of the remaining oil reserves are in Southern Russia, the Middle East, North Arica, and West Africa where stable governments are not necessarily the order of the day.

There will continue to be high risks of international and national terrorism, overthrow of existing governments and deliberate supply disruption in the years ahead. In addition, exporting nations may wish to keep their oil in the ground to maintain their target price range. In fact, there are considerable political and social uncertainties that could result in less oil being available than existing models predict.

5.4.4 Price Swings

The price of oil, like the price of all commodities, is subject to major swings over time, particularly tied to the overall business cycle. When demand for a commodity such as crude oil exceeds production capacity, the price will rise quite sharply because both demand and supply are fairly inelastic in the short run. Users of oil might be shocked by much higher prices, but they have commitments and habits that determine their energy use, and these take time to adjust.

On the supply side, especially at the outer edge of existing production capacity, adding new capacity is time-consuming and expensive. However, over time, both businesses and individuals figure out ways to cut back their oil consumption in response to high prices, and the high prices promote new investment in production and the arrival of new sources in the market, gradually restoring a supply-demand balance. The extraordinary spike in prices in mid-2008 represents to a large extent the consequences of a brief period where global oil demand outran supply. When supply exceeds demand, on the other hand, microeconomic theory says the price should collapse to the marginal cost of production of the most expensive source. As the price drops, the most expensive wells become uneconomical and are shut down, at least temporarily.

As global oil production begins to decline after peak oil, the medium-term volatility of oil prices is likely to be higher than before, because the range of production costs among all sources supplying the market will much greater. Major oil fields exist where the cost of production is relatively inexpensive and a large portion of the world's supply still comes from such inexpensive sources.

However, it is hoped that future shortages and high prices will spur the development of oil sources with higher production costs, including deep water sites, tar sands, oil shale, and secondary recovery from depleted fields. In the language of economic theory, the supply curve will be much steeper than in past years, and shifts in demand, either up or down, will cause relatively larger swings in market price.

5.5 References

BP. 2008. BP Statistical Review of World Energy. www.bp.com/statisticalreview.

Campbell, C.J. and Laherrère, J.H. 1998. The End of Cheap Oil. Scientific American. 278: 78–83.

Cleveland, C.J., Costanza, R., Hall, C.A.S., and Kaufmann, R. 1984. Energy and the United States Economy: A Biophysical Perspective. Science 225: 890–897.

Cobb, C., and Goldwhite, H. 1995. Creations of Fire: Chemistry's Lively History from Alchemy to the Atomic Age. Plenum Press, New York.

Cooper, C. and Pope, H. 1998. Dry Wells Belie Hope for Big Caspian Reserves. Wall Street Journal, 12 October.

Davis, L.J. 2003. Fleet Fire: Thomas Edison and the Pioneers of the Electric Revolution. Arcade Publishing, New York, 2003.

EIA. 2003. International Outlook 2003. Energy Information Administration, US Department of Energy, Washington, DC Report No. DOE/EIA-0484(2003). http://www.eia.doe.gov/oiaf/ieo/oil.html.

Forbes, R. J. 1958a. A History of Technology, Oxford University Press, Oxford, England.

Forbes, R.J. 1958b. Studies in Early Petroleum Chemistry. E. J. Brill, Leiden, The Netherlands.

Forbes, R.J. 1959. More Studies in Early Petroleum Chemistry. E.J. Brill, Leiden, The Netherlands.

Forbes, R. J. 1964. Studies in Ancient Technology. E. J. Brill, Leiden, The Netherlands.

Gingras, M., and Rokosh, D. 2004. A Brief Overview of the Geology of Heavy Oil, Bitumen and Oil Sand Deposits. Proceedings, Canadian Society of Exploration Geophysicists National Convention. http://www.cseg.ca/

Hakes. J. 2000. Long Term World Oil Supply. American Association of Petroleum Geochemists, New Orleans, Louisiana. http:www.eia.doe.gov/pub/oil_gas/petroleum/presentations/2000/long_term_supply/index.htm

Hall, C.A.S., Lindenberger, D., Kummel, R., Kroeger, T., and Eichhorn, W. 2001. The Need to Reintegrate the Natural Sciences with Economics. Bioscience, 51: 663–673.

Hall, C.A.S., Tharakan, P.J., Hallock, J., Cleveland, C., and Jefferson, M. 2003. Nature, 426: 318–322.

Hartshorn, J.E. 1962. Oil Companies and Governments: An Account Of The International Oil Industry In Its Political Environment. Faber and Faber, London.

Henry, J.T. 1873. The Early and Later History of Petroleum. Volumes I and II. APRP Co., Philadelphia, PA.

Hoiberg, A. J. 1964. Bituminous Materials: Asphalts, Tars, and Pitches. John Wiley & Sons. New York.

Hubbert, M.K. 1962. Energy Resources. Report to the Committee on Natural Resources, National Academy of Sciences, Washington, DC.

Kaufmann, R.K. 1991. Oil Production in the Lower 48 States: Reconciling Curve Fitting and Econometric Models. Res. Energy. 13: 111–127.

Kaufmann, R.K., and Cleveland, C.J. 2001. Oil Production in the Lower 48 States: Economic, Geological and Institutional Determinants. Energy J. 22: 27–49.

Lynch, M.C. 2002. Forecasting Oil Supply: Theory and Practice. Quarterly. Revs. Econ. Finance. 42: 373–389.

Munasinghe, M. 2002. The Sustainomics Trans-Disciplinary Meta-Framework for Making Development More Sustainable: Applications to Energy Issues. Int. J. Sustain. Dev. 5: 125–182.

Oil Fields. 2008. Oil Fields and What They Do (or Might) Produce and When: A List Of Major Oil Fields, Their Reserves, In Relation To Peak Oil. http://www.naturalhub.com/slweb/fading_of_the_oil_economy_oilfield_depletion_discovery_reserves.htm

Penrose, E.T. 1969. The International Petroleum Industry. MIT Press, Cambridge, Massachusetts.

Pirog, R. 2007. The Role of National Oil Companies in the International Oil Market August 21, 2007. Report No. RL34137. CRS Report for Congress, Congressional Research Service, Washington, DC. August 21.

Sadorsky, P. 1999. Oil price shocks and stock market activity. Energy Econ. 21: 449–469.

Sampson, A. 1975. The Seven Sisters - The Great Oil Companies and the World They Made. Hodder and Stoughton, New York.

Setright, L.J.K. 2003. Drive On! A Social History of the Motor Car. Granta Books, London, England.

Shafiq, 2009a. T. Iraq's Oil Prospects Face Political Impediments - 1. Oil & Gas Jounral, 107(3): 46–49.

Shafiq, 2009b. T. Iraq's Oil Prospects Face Political Impediments -2. Oil & Gas Jounral, 107(4): 31–36.

Smulders, S., and de Nooij, M. 2003. The Impact of Energy Conservation on Technology and Economic Growth. Resource Energy Econ. 25: 59–79.

Speight, J.G. 2007. The Chemistry and Technology of Petroleum, 4th Edition. CRC Press, Taylor & Francis Group, Boca Raton, Florida.

Speight, J.G. 2008. Handbook of Synthetic Fuels, McGraw-Hill, New York.

Speight, J.G., and Cockburn, S. 2008. Linkages of the Elements of the Oil and Gas Industry in Trinidad and Tobago with Other Industries in the

Economy. The Tobago Gas Technology Conference, Vanguard Hotel, Tobago, November 10–12.

Tainter, J. 1988. The Collapse of Complex Systems. Cambridge University Press, Cambridge, England.

TD Securities. 2007. Overview of Canada's Oil Sands.

Tharakan, P.J., Kroeger, T., and Hall, C.A.S. 2001. Twenty-Five Years of Industrial Development: A Study of Resource Use Rates and Macro-Efficiency Indicators for Five Asian Countries. Environ. Sci. Policy 4: 319–332.

USGS. 2000. United States Department of Long Term World Oil Supply. United States Geological Survey, Washington, DC. http://www.eia.doe.gov/pub/oil_gas/petroleum/presentations/2000/long_term_supply/index.htm

USGS. 2003. The World Petroleum Assessment 2000. United States Geological Survey, Washington DC. www.usgs.gov

Wilson, A. 1941. Persia – A Political Officer's Diary. Oxford University Press, London, England.

Yergin, D.H. 1991. *The Prize: The Epic Quest for Oil, Money, and Power.* Simon and Schuster, New York.

Zittel, W., and Schindler, J. 2007. Crude Oil: The Supply Outlook. EWG Series No. 3/2007, Energy Watch Group, Berlin, Germany. October.

6

Oil Prices

Petroleum economics is the field that studies human utilization of petroleum resources and the consequences of that utilization. In the simplest scientific terminology, petroleum use allows the production of energy. Resources can be viewed as renewable or depletable; petroleum falls into the latter category, which as an effect on pricing strategies. However, the projections of running out of oil are based on geology, not price. Most crude oil reservoirs have more than 50% of the original oil in place; many reservoirs have more than 60% of the original oil in place. These are resources that are known to exist. Much of the oil that is left is trapped in tiny pores and cannot be recovered by simple pumping, but requires more advanced and costly techniques covered under the umbrella of enhanced oil recovery (EOR) (Chapter 3).

As noted in previous chapters, petroleum is the energy source that dominated the 20th century, and barring unforeseen incidents, will continue to be a major source of energy for the first fifty years of the 21st century. In fact, petroleum is also the single largest commodity in international trade but price volatility shows the tenuous nature of the trading floor when petroleum is the focus of attention.

However, although petroleum is a viable and extremely versatile energy source, it is also the most political of energy sources not only because the resources are for the most part located in countries that do not require as much energy as the non-producing countries but because of the frailty of the various governments. In any of the oil-producing countries, governments can change remarkable quickly bringing different policies to bear on this natural resource. For this resource, countries have been willing to go to war, which is almost a catch-22 because it is a resource that is required for countries to wage war.

The price of a barrel of oil is highly dependent on both its grade, which is determined by factors such as its specific gravity or API and its sulfur content (Chapter 2). Location of the oil also plays a role in the ultimate price, and reference to the price of oil is usually either reference to the spot price of benchmark crude oil, such as West Texas (light) crude oil traded on New York Mercantile Exchange (NYMEX) for delivery in Cushing, Oklahoma; or the price of Brent (North Sea) crude oil traded on the International Petroleum Exchange (IPE) for delivery at Sullom Voe. On the other hand, the United States Energy Information Administration (EIA) uses the Imported Refiner Acquisition Cost (the weighted average cost of all oil imported into the United States) as the world oil price.

Crude oil prices have seen wide price swings over the past decade — especially over the past five years — whether it is due to apparent shortage or oversupply, (i.e. supply and demand factors) (ITF, 2008). At the time of writing, prices have seen highs approaching $150 per barrel and are currently touching $40 per barrel, which makes the average price (a number per barrel placed on the table by some economists) seem meaningless. While the average price may be meaningful to the statisticians in Washington DC, the real price of crude oil is another matter because of the real price of petroleum products, such as gasoline, may be an affordable $1.50 per gallon or an expensive $4.00 per gallon and not a stagnant, reasonable average of $2.75 per gallon. Even when adjusted for inflation to current dollars, an average price per barrel bears little relationship to reality, especially when the consumer cannot afford the products or must go into financial hardship because of the need for crude oil products.

To use an average price is to hide the truth of the barely or unaffordable higher prices. The analogy often used is a scientist or engineer standing with his left foot in a pail of boiling water and his right foot in a pail of ice water and declaring that he is comfortable because he is at average temperature.

In summary, the very long-term view of petroleum pricing can be considered in the same way. Average prices, even when adjusted for inflation, do not help the consumer who has to bear the brunt of the price increases. But, that is only part of the story — the rest follows.

6.1 Oil Price History

As the 20th century began, petroleum was being found, produced, and wasted. In the United States, the oil-producing states had to step into the production of petroleum to protect their resources. The relatively short producing life and resultant failure of the Spindletop field was one of the tragedies caused by a development that was based on haste, perhaps even on greed through application of the I-want-it-now scenario and the resulting too-much-too-soon scenario.

As the industry progressed in the United States, changes were necessary, and after World War II, the global nature of crude oil changed the supply structure. Historically, crude oil prices have varied from $2.50 to about $3.00 in 1957 to almost $150 per barrel in the late summer of 2008. In the early part of this time period, the oil-producing countries found increasing demand for their crude oil but, as US demand increased and foreign supplies of petroleum became available, prices were largely defined by what refineries, usually owned by the larger oil companies, were willing to pay. This system worked fine for refineries but not for the producers.

In 1972 the price of crude oil was about $3.00 per barrel, and prior to October 1973, the world oil suppliers had very little say in setting the oil prices — the oil importers (i.e., the oil companies) set the prices. As a result of the Arab-Israeli war of 1973, oil became the weapon of political pressure and the OPEC member nations became assertive and organized; oil prices increased to $12 a barrel in 1974. The United States went into a major economic recession from 1981 to 1983 and the demand for oil was reduced and oil-producing nations kept pumping oil in competition to each other to maximize cash inflow. As a result, oil prices decreased, and in 1985, prices were again on the order of $13 — close to the price as in 1974.

The crude-oil price spike of the early 1970s was due to an embargo imposed by the Organization of Petroleum Exporting Countries (OPEC). Less oil without an immediate corresponding decline in oil demand drove prices up. Even if businesses could pay the higher price, they could not get the oil they needed to sustain the pace of economic activity. As a result, real output slowed and costs rose. This put pressure on profit margins, which in turn forced businesses to boost prices. Meanwhile, workers expected prices to continue to rise and demanded even higher wages. Without corresponding productivity gains, this led to stagflation later in the decade (Laufenberg, 2007).

In mid-1985, oil prices were linked to the spot market for crude, and by early 1986, prices decreased to the lowest levels since the early 1970s and moved downward to $8 to $10 per barrel. The price of crude oil rose again in 1990 with the Iraqi invasion of Kuwait and the ensuing Gulf War, but following the cessation of hostilities, crude oil prices entered a steady decline. However, since then, the demand in the United States and Europe has increased approximately 6 to 8% a year, which has, at times, been a strain on crude oil supply.

From 1990 to 1997, world oil consumption increased by more than six million barrels per day, but the price increases came to an end when, due to downward trends in several Asian economies, higher OPEC production sent prices downward. In late 1997 several events combined to initiate a precipitous drop in world oil prices:

1. Asian economies, which had been generating the greatest increases in petroleum demand, suffered substantial contractions causing a lowering petroleum use.
2. The OPEC member nations, who may have mis-read this situation, agreed to increase oil production.
3. The Northern Hemisphere benefited from a mild winter and the demand for crude oil and crude oil products was reduced.
4. Weakness in the Russian economy resulted in higher exports of Russian petroleum.
5. Venezuela and Saudi Arabia engaged in a market share battle that led to higher volumes of petroleum exports.

As a result of the First Gulf War, petroleum production by Iraq was defined by the United Nations' sanctions program. In addition, the United Nations allowed Iraq to increase the amount of oil it could produce and sell; at the beginning of 1998, Iraq exported approximately 500,000 barrels/day of crude oil, but by the beginning of 1999 Iraq was exporting 2.5 million barrels/day. At the same time, other OPEC countries were reducing production as a means of controlling oil prices, but this was offset by the increases in Iraqi crude oil production and oil prices were on a downward slope.

In March 1999, the OPEC members agreed to reduce exports to the oil consuming countries; at this time Mexico, Norway, and other (until then non-OPEC) oil-producing countries joined the OPEC cartel. As demand increased, prices were affected, and by the end of 1999, oil prices had returned to 1997 levels. The consequences of price reductions were obvious in a net loss of upstream jobs and oil rigs no longer in use. From 1999 onwards, new influences entered the oil price market. China and India added on to the market demand as importers. India had a rapidly increasing economy thereby placing increasing the demand for oil and, by 2000, prices had increased to approximately $27 per barrel.

In March 2000, OPEC acted to increase production but when the OPEC members had agreed to cut production Saudi Arabia agreed to the biggest individual reduction to offset the increased production share that Iraq had acquired. However, when increases in production were being considered, no OPEC member country wanted to relinquish market share in favor of Saudi Arabia, but many of the member countries had now lost their previous production capacity.

OPEC members began to try to control production to keep crude oil within a price band ranging from $22 per barrel to $28 per barrel. This range is believed to reflect a balance that provides the OPEC member nations with the income necessary to meet individual national budgets while maintaining an acceptable market price. However, at the beginning of 2001, the strain on worldwide production capacity caused prices to exceed $30 per barrel, but conservation and the developing recession in the United States began to drive demand down with an ensuing decrease in oil prices. As a result, the OPEC member countries initiated production cuts but the commitment to meet quotas had waned and targets were not met.

At the end of 2001, crude oil prices saw a steady increase and reached $40 to $50 per barrel by September 2004. In October 2004, the price of crude oil exceeded $53 per barrel and for December delivery exceeded $55 per barrel. Crude oil prices surged to a then record high above $60 a barrel in June 2005, and to the unheard of prices in excess of $145 per barrel in the late summer of 2008.

Historically, a recession may be expected to follow a spike in oil prices but higher crude-oil prices alone are not enough to cause a recession and other factors must be considered (Laufenberg, 2007).

The key is to not simply acknowledge that the price of crude oil is higher, but rather ask *why*?

6.2 Pricing Strategies

OPEC was formed in 1960 with five founding members: Iran, Iraq, Kuwait, Saudi Arabia, and Venezuela. By the end of 1971, six other nations (Qatar, Indonesia, Libya, United Arab Emirates, Algeria, and Nigeria) had joined of OPEC. By 1973, OPEC controlled enough petroleum production that, when acting collectively, it could determine whether the world oil supply and could subsequently define the market price.

Ultimately, the OPEC member nations set new production quotas and, at the same time, a change in petroleum pricing was instituted. In 1983, the New York Mercantile Exchange (NYMEX) began to trade crude oil futures on the commodity market. This meant that commodity market trading would become the price maker and crude oil prices would be set on the trading floors of the NYMEX that, in turn, is set by supply and demand, with the oil-producing countries controlling demand. But, this is a double-edged sword.

However, Russia, one of the major oil producing countries, has become a key to the course of future crude oil supply politics. Since January 2000, Russian crude oil production has increased by more than 900,000 barrels per day, which has offset the reductions made by OPEC to stabilize crude oil supply. Without significant action by Russia to reduce its production, there is the risk of an oil price war.

Oil prices contribute significantly to the national budgets of oil-producing countries. A high oil price encourages prosperity in such countries. Whether or not the oil-consuming countries move to other sources of energy and learn to control the demand for oil, crude oil prices and crude oil production may be lowered and prosperity will suffer. This has the potential to destabilize many of the governments of the oil-producing countries. The result is possible anarchy through the move to a series of unstable or fickle governments, which may send prices into an upward spiral once more.

Nevertheless, over the last three decades, other than token efforts, the oil-consuming nations have not launched into alternate energy programs in a manner sufficient to influence crude oil prices. It is the swings in the price of crude oil that have controlled the alternate-energy-related programs of the oil-consuming nations. Over the past decades, not one country has made the decision to bite the bullet and initiate full-scale research into alternate energy with the goal of commercialization. On the other hand, the OPEC member nations know it is a matter of economics that to conserve the sales of a product, the price of that product must remain competitive. If oil is to remain competitive, the price must be such that it will be uneconomical for oil-consuming nations to institute programs related to alternate energy. In 1978, Mr. Sadam Hussein (then-Vice President of Iraq) explained this to the author very simply and effectively.

However, for the purposes of pricing, crude oil is generally classified based on the API gravity and sulfur content (Chapter 4). For example, light sweet crude oil has low density, low viscosity (there are no exact numbers assigned to this, because the classification is more practical and theoretical), and low sulfur content making it easier to transport and refine and, therefore, more expensive to purchase. For example, the petroleum industry generally classifies crude oil by the geographic location it is produced in (e.g. West Texas, Brent, or Oman), its API gravity, and by its sulfur content.

Light sweet (high API gravity) crude oil is more desirable than heavy sour (low API gravity) crude oil because it produces a higher yield of gasoline (Table 6.1), while sweet oil commands a higher price than sour oil because it has fewer environmental problems and requires less refining to meet sulfur standards imposed on fuels in consuming countries. Each crude oil has unique molecular characteristics that are understood by the use of crude oil assay analysis in petroleum laboratories. Some of the common reference crudes are (Chapter 4):

- West Texas Intermediate (WTI), a very high-quality, sweet, light oil delivered at Cushing, Oklahoma for North American oil
- Brent Blend, comprising 15 oils from fields in the Brent and Ninian systems in the East Shetland Basin of the

North Sea. The oil is landed at Sullom Voe terminal in the Shetland Isles. Oil production from Europe, Africa and Middle Eastern oil flowing west tends to be priced off the price of this oil, which forms a benchmark
- Dubai-Oman, used as benchmark for Middle East sour crude oil flowing to the Asia-Pacific region
- Tapis (from Malaysia, used as a reference for light Far East oil)
- Minas (from Indonesia, used as a reference for heavy Far East oil)
- The OPEC Reference Basket, a weighted average of oil blends from various OPEC (The Organization of the Petroleum Exporting Countries) countries.

There are declining amounts of these benchmark oils being produced each year, so other oils are more commonly delivered. While the reference price may be for West Texas Intermediate delivered at Cushing, the actual oil being traded may be discounted Canadian heavy oil delivered at Hardisty, Alberta, and for a Brent Blend delivered at the Shetlands, it may be a Russian Export Blend delivered at the port of Primorsk.

Table 6.1 Comparison between assays of light sweet crude oil and heavy sour crude oil.

Liquid Volume %	Light Sweet	Heavy Sour
Gas (Initial Boiling Point to 99°F)	4.40	3.40
Straight Run (99 to 210°F)	6.50	4.10
Naphtha (210 to 380°F)	18.60	9.10
Kerosene (380 to 510°F)	13.80	9.20
Distillate (510 to 725°F)	32.40	19.30
Gas Oil (725 to 1050°F)	19.60	26.50
1050 + Residuals	4.70	28.40
Sulfur %	0.30	4.90
API	34.80	22.00

Not only is the price of crude oil subject to composition of the crude, but the price of crude oil is combination of the following factors:

1. The price set by the producer nations
2. Continued high demand in the industrialized world as seen in the increase in gas-guzzling vehicles such as the SUV
3. Increased car ownership in developing world where India and China are rapidly developing nations with need for petroleum and petroleum products, and
4. A host of international factors such as the Second Gulf War, and US-Iran tensions, to name only two of the last category.

Each of these has contributed significantly to the price of crude oil, and the United States, rightly or wrongly, has contributed to almost all of the above issues with soaring demand, invading Iraq, and raising tension with Iran.

Crude oil price has been hostage to the world events. Also, production cuts or restoration by OPEC gravely impact the market place. Since the year 2000, the following political events impacted the oil prices:

- September 11 2001 attack the World Trade Center
- Gulf War II begins in early 2003
- Insurgency in Iraq (2003 to present)
- Iranian nuclear debacle started in 2005
- Russian entry into the world market from year 2000 onwards
- Hurricane season of year 2004 in the US
- Hurricane Katrina and Rita in the year 2005
- Rising demand, which supply could not meet.

All of the above have played their role in the per-barrel price. Some events have impacted more significantly than the others. The net impact has been that the consumers of crude oil subject to any one of several events that can influence the price. However, notable among the world events that have affected oil prices have been the Gulf War II, 2005 Hurricane season, and Iran-US Nuclear tension. Although the 9/11 can be classified in this category, oil prices merged from that disastrous event almost unchanged. In fact, oil demand dipped immediately after the 9/11 attack, resulting in a production cut by OPEC.

The Gulf War II and the insurgency in Iraq cut off oil supply from Iraq. This caused a shortage in supply and, hence, in a rise in price. By the third quarter of 2005, the price of oil had increased to $45 a barrel. Later, this price increased further and then momentarily stabilized at approximately $50 per barrel.

In the late summer of 2005, hurricanes Katrina and Rita, which hit the Gulf Coast of the United States, had a substantial effect on the price of oil. A number of oil refineries shut down and number of oil production platforms damaged. Crude oil rose to $60 per barrel.

The Iran-US Nuclear tension created international tension as peaceful means to ask the Iranians to cease and desist failed. The decision of the Bush administration to invade Iraq made a military option (involving a strike to cripple Iran's nuclear facilities) a distinct possibility. This, accompanied by the crippled production at the US Gulf Coast and the management of supply to demand by OPEC, started the further upward movement of oil prices.

6.3 Oil Price and Analysis

Crude oil is one of the main natural feedstocks used to meet energy demands of mankind and price variation has a significant influence on the society development. The prognoses of the volume of crude oil extraction globally and regionally, consumption rates, and the crude oil price are used not only for planning the national and world economies, but also for development of refining enterprise investment programs.

Despite some prognoses for depletion of crude oil reserves in the near future, the structure of energy consumption is not expected to significantly change. A marginal reduction of the crude oil influence till 2030 (up to 32%) and equalization of the energy balance of crude oil, natural gas and coal is expected up to the mid-point of the 21st century.

One of the main advantages of the crude oil as an energy source is its application in the transport sector where the oil-refined products are and keep on being without serious competition. As the population around the globe becomes more mobile, as is currently happening in India and China, the demand for liquid fuels in the transport sector will increase in contrast to the reduction of energy consumption in the other sectors.

Crude oil prices behave much as any other commodity with wide price swings in times of shortage or oversupply. The crude oil price cycle may extend over several years responding to changes in demand as well as OPEC and non-OPEC supply. However, in the United States, the price of crude oil has been heavily regulated through production or price controls throughout much of the 20th century.

The oil industry in the United States has been subjected to varying degrees of price controls since August 1971, when general price controls were levied on the entire US Economy. As controls were phased-out in other industries, more stringent price regulations were imposed on the oil industry in response to the October 1973 oil embargo and the subsequent quadrupling of world oil prices (Helbling and Turley, 1975).

The oil price control program is directed at cushioning the domestic impact of sharply higher external oil prices. In this respect, the controls effort can be regarded as successful because the effective domestic price for petroleum remains, in fact, below world market prices. Economic analysis, however, indicates that the controls will (1) become ineffective, over time, with respect to the above stated intention and (2) will enhance the ability of external suppliers to manipulate prices.

Using economic theory as a foundation, the eventual effects of controls on domestic production, imports, and the domestic price of oil are derived. In this regard, two of the more popular concepts are that decontrol will result in (1) higher domestic petroleum prices and (2) increased domestic production and reduced imports?

United States oil refiners currently process about 18 million barrels per day. Of this total, approximately 12 million barrels per day are imported from other countries. The United States did not always rely to such an extent on external oil supplies. In the mid-1960s, oil imports represented only 20% of total U. S. Consumption. In fact, as late as 1971, import quotas on petroleum products existed in order to prevent relatively cheap foreign oil from placing domestic oil producers at a competitive disadvantage.

In the months of July until December 31 2008, oil prices fell from $147 a barrel to less than $40 per barrel. However, since 2003, even in a strong upward trend of oil prices, there have been several dips in the price ranging from 10% to 31%. So, the latest retreat of 22% should not be so unexpected.

Notwithstanding, many saw this 2008 price retreat as a continuing stable trend to much lower levels and rejoiced over the permanent relief that much cheaper oil would bring. However, it is felt that as oil stocks are depleted further and peak oil becomes a reality, liquid fuel prices and price volatility will increase dramatically. With price volatility, there will be a dramatic shift between high prices and low prices and that even if the low prices are significantly lower than the high prices, this should not be used as a reason to withdraw from the investigation of alternate sources of liquid fuels. As with any commodity, crude oil prices have retreated and advanced with overall effect (in recent years) to move to new record prices. Thus, the likely overall trend for the price of crude oil will be aggressive price rises from a variety of underlying causes, which are not reflective of the price variations of many other commodities. Indeed, world political and geopolitical events, and economic growth and decline, have all influenced the price of oil over the decades.

Prior to 1973, the oil-producing countries had initiated two previous embargoes — one in 1956 and another in 1967. But, because the United States was an oil exporter, these embargoes had no affect on the United States. After 1970, the United States was an overall oil importer and it is not surprising that the oil-producers embargo of 1973 had a much greater effect on the United States. At the time, Saudi Arabia was assuming the role of price-controller through control of the production of oil. In the 1980s, as the North Sea and North Slope oil came on line, there was some effect on Saudi Arabia's control, but the favorable economics of producing oil from these two fields was not to last. The cost of oil production from the Saudi fields dropped below the cost of producing North Sea oil and price was again set by Saudi oil production. This was reinforced during the First Gulf War when Saudi Arabia more than made up for the lost of Kuwaiti and Iraqi oil production. However, since 2000, Saud Arabia no longer has the excess oil production capability to drive down the price of oil. Neither does any other producer, although Venezuela would like to move into the position of swing producer but, other than heavy oil resources, does not have the resources of conventional oil to do this.

Thus, the world, particularly the United States, is in a period where no one country has control over production and thus no one country has control over price.

Unless the Congress of the United States Government offers some other alternative to imported oil, crude oil prices will lie outside of the sphere of influence of the United States and there will be price volatile encompassing the variations between extremely low prices and extremely high prices.

The high price of oil is not palatable or affordable to many Americans. When the choice comes to a high price or a volatile price régime where the volatile high price of oil could exceed the stable high price, one wonders where the choice of the people might lie.

Finally, price clustering occurs when transaction prices are not evenly distributed among all possible ending values, but tend to cluster around even numbers. Price clustering may help traders simplify negotiations and it may also be an indication of market quality. Stock prices cluster on round fractions and price clustering rises with price level and volatility and declines with capitalization and transaction frequency (Harris, 1991; (Pirrong, 1996; Tse and Zabotina, 2001).

For related products trading in multiple venues, price discovery is essential in determining the dominant market in terms of the development of the asset's implicit efficient price. Hence, prices of regular-size NYMEX futures, NYMEX e-mini futures, WTI ICE electronic futures, and the USO ETF are strongly related and driven by the same information.

The prices of the crude oil futures and the USO ETF in different markets are kept from drifting apart because of inter-market arbitrage. These prices are therefore co-integrated and share one common factor — the implicit efficient price in which the concept of information shares (IS) is defined the relative contribution of each market to price discovery (Hasbrouck, 1995). However, this does not conclude which market has the best price but, instead, which market moves first in the price-correction process (Hasbrouck, 2002; Sapp, 2002). In fact, the information share of a trading market is the portion of the variance of the common factor that is attributable to innovations in that market and different ordering of the variables will produce lower and upper bounds of the information shares. Nonetheless, many other studies using lower frequency data present considerable differences between lower and upper bounds (Martens, 1998; Huang, 2002; Booth *et al.*, 2002). In fact, the average of the information shares given by all orderings is a reasonable estimate of the market contribution to the price discovery. Similar to earlier research, we discover a wide range between lower and upper information share estimations (Baillie *et al.*, 2002).

6.4 The Anatomy of Crude Oil Prices

Crude oil is one of the main natural feedstocks used to meet energy demands of mankind, and, as a result, price variation has a substantial influence on the society development. The prognoses of the global and regional volumes of crude oil production, consumption rates, and the crude oil price are used not only for planning the national and world economies, but also for development of refining methodology.

Since the beginning of this century, the energy consumption pattern has been almost unchanged and has involved use of crude oil (39.3%), natural gas (22.6%), coal (20.8%), renewable sources (3.9%), hydropower stations) (1.9%), and nuclear energy (10.6%) (IEA, 2008). Despite some prognostications for the depletion of crude oil reserves in the near future, the structure of energy consumption is not expected to significantly change. A marginal reduction of the role of crude oil in energy generation till 2030 and equalization of the energy balance of crude oil, natural gas and coal is likely until the mid-point of the 21st century.

As the pain induced by higher oil prices spreads to an ever growing share of the American and world population, pundits and politicians have been quick to blame assorted villains: oil companies, commodity speculators, OPEC, and domestic policies of the United States government. While each of these parties has contributed to and benefited from the increased prices, the sharp growth in petroleum costs is due far more to a combination of soaring international demand and slackening supply.

Many consumers believe that OPEC member states restrain crude oil production, and even though international oil markets efficiently price and allocate the crude oil being produced, some economists believe that the amount of crude oil being produced is a function of market power and that this exercise of market power produces greatly inflates world crude oil prices. For example, the Middle East with its vast reserves (65% of the world total) and highly prolific oil wells could have developed reserves to produce and sell enough oil to satisfy total world demand at under $5 per barrel and still enjoy substantial government revenues.

In the case of the current price oil fluctuations, another factor was emerging. Energy demand in mature industrial nations was continuing to grow as energy demand in India and China was also beginning to make an impact. By 2002, the United States Department of Energy was predicting that China would soon overtake Japan and become the world's second-largest petroleum consumer and that developing Asia as a whole would account for about one-fourth of global consumption by 2020. Also evident was an unmistakable slowdown in the growth of world production, the telltale sign of an imminent approach of an *oil peak* in global output (Klare, 2007).

With these trends in mind, it is again time to revisit policies that will minimize future reliance on oil in the United States. Energy conservation needs to be emphasized, and the development of climate-friendly, alternative sources of energy, such as biofuels, wind, solar, and geothermal, need to be promoted and followed assiduously. However, there are still those who believe that while energy conservation may be a sign of personal virtue, it is not a sufficient basis for a sound and comprehensive energy policy. During this time of investment in other sources of energy, there is still the need to rely on oil, natural gas, and coal.

Continued reliance on oil, of course, means increased reliance on imported petroleum, especially from the Middle East, but there needs to be a careful watch on the implications of continued dependence on imported oil, because oil from other domestic sources, such as the Arctic National Wildlife Refuge (ANWR) and previously-prohibited offshore areas, may only reduce the need for imports into the United States by a low percent.

Indeed, increased reliance on imports means increased vulnerability to disruptions in delivery due to wars and political upheavals. Increased military involvement in major overseas oil zones, especially the Persian Gulf, is not the answer. Furthermore, threats to countries that do not see eye-to-eye with the policies of the United States can also cause disruption in the supply of crude oil. But, such decisions cost the American people credibility and trust by foreign nations and present the picture of the United States as an unreliable trading partner.

Such actions usually initiate a rise in the price of crude oil, and while there is always the argument that the rise in the price of crude oil is due largely to increasing demand chasing insufficiently expanding supply, energy policies can intensified the problem.

While other sources of oil, such as the Arctic National Wildlife Refuge and offshore source, will not reverse the long-term decline in oil production in the United States, it is only by reducing demand that fundamental market forces can be modified. This is best done through a comprehensive program of energy conservation, expanding public transit and accelerating development of energy alternatives. And, we must not forget that the development of new technologies to recover oil from shut-in well is also necessary.

The ready availability of futures, spot, and contract markets suggests that market prices accurately reflect international supply and demand for crude oil. But, many believe that OPEC member states restrain crude oil production. Even though international oil markets efficiently price and allocate the crude oil being produced, most (but not by any means all) economists believe that the amount of crude oil being produced is a function of market power and that this exercise of market power produces greatly inflates world crude oil prices. For instance, Francisco Parra, former Secretary-General of OPEC, maintains that the Middle East with its vast reserves (65% of the world total) and highly prolific oil wells could have, if it had been so minded, developed reserves to produce and sell enough oil to satisfy total world demand at under $5 per barrel, and still enjoy substantial government revenues.

If the OPEC cartel does raise world crude oil prices by constraining production, are price controls warranted? From an economic perspective, the answer is no. Domestic price controls will not reduce OPEC's market power. The manner in which domestic price controls were implemented in the U.S. in the 1970s actually increased the demand for OPEC imports and thus, its profits and punished domestic producers who are not at fault for OPEC production decisions. Price controls also reduce incentives to increase production, whether OPEC is strangling the market or not. Domestic price controls assist the cartel's attempts to restrict supply.

Predictions of crude oil consumption and price rises have been made since the middle of last century. Assessment of the possible long-term trends in crude oil price variation should be made as the price variation is analyzed for sufficiently long period of time. Transitory variations, which have little significance, may be omitted. However, it is expedient to investigate (1) the reason for the step wise change of crude oil price, and (2) the main factors that affect crude oil price, and the anticipated.

The first accelerated increase of crude oil price has started immediately in the beginning of the Second World War. After the war developing industrialization and agricultural activities required more fuels and crude oil for agriculture activities and transport of goods from manufacturer to the consumer, which was followed by marginal price increases for two decades (1950–1970), allowing the new way to develop. Since then, the key drivers from increased energy demand and the accelerated increase of crude oil prices have been growing population and increased economic. Streamlining of energy consumption efficiency alleviates this factor, improving transport engine efficiency and the efficiency of management of the whole transport section). In addition, the increased cost of finding and developing oil is also a contributor to increased crude oil prices. For example, in 2006 it was estimated that the cost of finding and developing oil on a per barrel basis was three times greater than in 1999 (Gonzalez, 2006).

6.5 The Anatomy of Gasoline Prices

Gasoline accounts for about 50% of the consumption of petroleum products in the United States and the price of gasoline is the most visible among these products. As such, changes in gasoline prices are always under public scrutiny and price shocks can originate at any point from crude oil prices to the final price at the gasoline pump. Shocks originating at an intermediate step, such as the wholesale price of gasoline, may reflect a bottleneck in distribution, while price shocks originating farther upstream are more likely to represent the effects of variation in crude oil supply. Price shocks originating at the retail level are more likely to represent variation in the demand for gasoline. Given the history of oil-supply shocks and indications that demand for gasoline is relatively stable, intuition suggests that price shocks are more likely to originate upstream and be transmitted downstream (Norman and Shin, 1991; Balke *et al.*, 1998).

Recent price increases have led some to call for federal price controls for gasoline and/or related oil products as well as some form of windfall profits tax on the oil industry.

Proponents of intervention contend that gasoline markets are not competitive (with some accusing producers of price collusion), that fat profit margins induce little more supply than might otherwise be induced by healthy but reasonable profit margins, and that the gasoline profits are largely unanticipated and unearned. As a result, oil companies are reaping very large profits at the expense of consumers, and maintain that price controls and/or windfall profit taxes would simply redistribute wealth from producers to consumers without any significant effect on supply.

However, government intervention may improve overall economic efficiency if prices do not reflect total costs or if the market in question is not competitive. No matter how imperfect markets may be, government intervention poses new problems. Accordingly, evidence that market imperfections exist is a necessary, but not sufficient condition for government intervention.

In gasoline markets, no evidence supports any market failure claims in a manner that would support reduction of gasoline prices. For example, the social costs associated with gasoline consumption that are not fully reflected in the price of gasoline at the pump, but the implication is that market prices for gasoline, compared to the prices in many oil importing countries, are too low — not too high.

However, laws prohibiting retail gasoline outlets from pricing gasoline below cost, such as a mandatory minimum markup above a legally defined wholesale price, exist in many states. Several other states have more general minimum mark-up laws that pertain to gasoline as well as other products while other states prohibit vertically integrated oil companies from owning retail gasoline outlets. The intended effect of such laws is to keep some entrants out of the market such as (1) those companies that may sell gasoline at or near acquisition cost in order to encourage traffic and thus sales of other more profitable products, and (2) those companies that may undercut the prices charged by independent retail operators. This is to the detriment of gasoline consumers.

But crude and gasoline prices can diverge even in perfectly competitive gasoline markets. The temporary increase of gasoline prices following Hurricane Katrina illustrates this point. Approximately 2 million barrels of refining capacity a day (approximately 11% of total refining capacity in the United States) were shut down as a consequence of the storm causing a disruption in fuel delivery from Gulf Coast refineries. The supply of gasoline at retail outlets greatly decreased and, hence, increased retail prices beyond what might otherwise have been expected from the overall 2% decrease in world crude oil production as a result of the storm. Furthermore,

Hurricane Rita reduced additional refining capacity causing a shutdown of 1 million barrels of gasoline production per day.

Analysis of retail gasoline pricing is complicated by large price differences in various parts of the country. In fact, the phenomenon of prices at different market levels tending to move differently relative to each other depending on direction is known as price asymmetry. To the public, this typically means simply the notion that retail prices rise faster than they fall, when observed over some period. However, there is significantly more to the question of price asymmetry than just the upward and downward speed of retail price movements. For the most part, retail prices move in response to changes in wholesale, or even raw material, prices further upstream in the manufacturing/distribution chain. Therefore, an examination of price asymmetry must consider the speed and degree to which price changes at one level are passed downstream (i.e. from wholesale toward retail). In previous studies of this phenomenon, researchers have further defined two types of price asymmetry: amount asymmetry, in which the amount of the eventual price change differs between wholesale and retail and/or between upward and downward movements, and pattern asymmetry, in which the change occurs at a different rate between market levels depending on direction.

There are two key concepts to keep in mind when analyzing a gasoline market: asymmetry and pass-through. As noted above, asymmetry refers to prices rising and falling at differing rates at different levels in the pricing structure. However, the analysis is complicated by the fact that there are lags between changes in upstream prices and the corresponding changes in downstream prices. The upstream price changes take time to pass-through to the downstream prices. The pass-through times make it theoretically possible for there to be no real asymmetry in price movements, but because of the time lags a statistical test may show that there is asymmetry.

For example, a very simple price change, such as a symmetrical $0.5 per gallon upstream price increase and decrease spread over 10 weeks, can have unusual consequences because of lags. The downstream price peaks after the upstream price maximum, and it takes longer for the downstream price to return to equilibrium than the upstream price. Lags anywhere in the system can therefore give the appearance that prices are sticky downward, when in fact the apparent asymmetry is only an artifact of the lags.

The terms upstream prices and downstream prices are used to distinguish the relative position of two prices in the distribution system. Upstream prices are commodity prices closer to the production point relative to downstream prices, which are motor gasoline prices closer to the final consumption point. For example, retail gasoline prices are downstream relative to all other commodity prices because they constitute the price the charged to the consumer. Also, pipeline spot gasoline prices are downstream from United States Gulf Coast spot gasoline prices because they are closer to the consumer. Conversely, United States Gulf Coast USG spot prices are upstream relative to pipeline spot prices because the United States Gulf Coast spot prices are further removed from the final consumption of gasoline and are closer to the production side. Crude oil spot prices are upstream of all motor gasoline prices because they are farthest from the final consumer.

6.6 Effect of Refining Capacity

The demand for refined products, particularly transportation fuels, has been rising steadily over the past three. As demand for petroleum products has grown, their quality and performance have changed substantially as a result of environmental regulations and motor vehicle performance requirements. The refining industry is already upgrading facilities to produce ultra-low-sulfur gasoline and diesel fuel and to reduce the environmental impacts of plant operations.

The 1990s were widely viewed by the industry as a period of economic volatility that was characterized by poor profit margins as a result of the increasing cost of compliance with environmental regulations and unfavorable crude oil price trends. At the same time, the corporate restructuring and consolidation has dramatically changed the refining industry in the United States (Peterson and Mahnovski, 2003). In addition, the crude oil slate into refineries was changing to include low API gravity/high sulfur crude oil (Swain, 1991, 1993, 1997, 2000) and operations were highly dependent upon the ability of current refinery operations to accept crude oil for production of the different products (Speight, 2007).

Different types of crude oil yield a different mix of products depending on the qualities of crude oil. Crude oil types are typically differentiated by their density (measured as API gravity), as well as their sulfur content. Crude oil with a low API gravity is considered a heavy crude oil, and typically has higher sulfur content and a larger yield of lower-valued products. Therefore, the lower the API of a crude oil, the lower the value it has to a refiner as it will either require more processing or yield a higher percentage of lower-valued by-products, such as heavy fuel oil, which usually sells for less than crude oil.

6.6.1 Refinery Types and Crude Slate

A refinery is an installation that manufactures finished petroleum products from crude oil, unfinished oils, natural gas liquids, and other hydrocarbons using heat, pressure, catalyst, and chemicals. Early refineries were simple, batch distillation units that processed hundreds of barrels of crude oil per day from one or a small collection of fields. Current refineries are complex, highly integrated facilities that contain a dozen or more process units capable of handling numerous crude oils from around the world with capacities generally varying from 100,000 to 500,000 barrels per day.

Every crude oil produced in the world has a unique chemical composition that will determine the optimal manner in which to process the crude and the final product slate. Crude oil contains distillates of different molecular composition and shape, burning qualities, and impurities, such as metals, asphaltene constituents, nitrogen, and sulfur (Speight, 2007). The types, size, number, and flow sequences for each refinery vary, depending on crude oil input quality, output product slate and quality, and environmental, safety, and economic conditions.

Refineries evolve with changes in market demand, feedstock, product specification, and environmental regulation; and will typically possess both very old and modern process units. In most refineries, operating variables, such as temperature, pressure, residence time, feed quality, cut points, recycle-gas ratio, space velocity, and catalyst allow refineries to balance feedstock, product, and quality. The limits of these variables are unique to the plant design and nature of the input and output requirements. For a specific feedstock and catalyst package, the degree of processing and conversion increases with the severity of the operation. The type of processing units at the refinery will influence a refiner's choice of crude oil. Refineries fall into four broad categories:

1. Topping refinery
2. Hydroskimming refinery
3. Cracking refinery
4. Coking refinery.

The topping refinery is the simplest refinery and consists only of a distillation unit and probably a catalytic reformer to provide octane. Typically only condensates or light sweet crude would be processed at this type of facility unless markets for heavy fuel oil are readily and economically available. Asphalt plants are topping refineries that accept heavy crude oil from which the asphalt is produced.

The next level of refining is the hydroskimming refinery, which is equipped with a distillation unit as well as naphtha reforming and necessary treating processes. The hydroskimming refinery is more complex than the topping refinery and it produces gasoline.

The cracking refinery is used to process the gas oil fraction from the crude distillation unit (a stream higher boiling than diesel fuel but lower boiling than heavy fuel oil) and converts it to gasoline and distillate components using catalysts, high temperature, and high pressure.

The last level of refining is the coking refinery. This refinery processes residual fuel, the heaviest material from the crude unit and thermally cracks it into lighter product in a coker or a hydrocracker. The addition of a fluid catalytic cracking (FCC) unit or a hydrocracker significantly increases the yield of higher-valued products like gasoline and diesel oil from a barrel of crude, allowing a refinery to process cheaper, heavier crude, while producing an equivalent or greater volume of high-valued products. Hydrotreating is a process used to remove sulfur from finished products. As the requirement to produce ultra low sulfur products increases, additional hydrotreating capability is being added to refineries. Refineries that have large hydrotreating capability have the ability to process crude oil with higher sulfur content.

The process technologies in a coking refinery vary from distillation to coking, catalytic cracking, catalytic reforming, catalytic hydrocracking, catalytic hydrotreating, alkylation, polymerization, aromatics extraction, isomerization, oxygenates, and hydrogen production (Gary *et al.* 2007, Speight 2007). In addition, the crude slate is expected to change significantly in the years ahead as refiners increase their capacity to process heavy crude oil and lower quality synthetic crudes. Refineries that are dependent on imported crude oil tend to process a more diverse crude slate than their counterparts where the oil is shipped from a domestic wellhead. These refiners have the capacity to purchase crude oil produced almost anywhere in the world and therefore have flexibility in their crude buying decisions.

As supplies of light sweet crude oil continue to deplete, refiners will increasingly turn to heavy sour crude. But, not enough refiners yet have a demand for heavy sour, so it trades at a significant discount to light sweet crude oil. This will of course change as more coking units are installed in refineries. There will be a higher demand for heavy crude oil, and the asphalt market will become more lucrative as the atmospheric and vacuum residua are routed to coking units to increase liquids production. Thus, economics currently favor installing coking units and hydrotreaters to handle the heavy sour crudes, and will continue to do so as long as they trade at a substantial discount to light sweet crudes.

The installation of additional conversion capability increases the yield of clean products and reduces the yield of heavy fuel oil. However, increased conversion capability would generally result in higher energy use and, therefore, higher operating costs. These higher operating and capital costs must be weighed against the lower cost of the heavier crude oil. With increased heavy oil production and the declining production of conventional light sweet crude oil, many refineries have undertaken the investment required to process the increasing supply of heavy crude oil. Much of this investment is by the large integrated oil companies — those that are involved in both the production of crude oil and the manufacturing and distribution of petroleum products.

Thus, refinery configuration is also influenced by the product demand. For example, in the United States, the demand for gasoline is much larger than distillate demand and, therefore, refiners configure their installations to maximize gasoline production. On the other hand, in several Western European countries, most notably Germany and France, policies exist that encourage the use of diesel engines creating a much stronger distillate component.

The relationship between gasoline and distillate sales can create challenges for refiners. A refinery has a limited range of flexibility in setting the gasoline to distillate production ratio. Beyond a certain point, distillate production can only be increased by also increasing gasoline production.

6.6.2 U.S. Refining Capacity

The refining capacity in the United States, as measured by daily processing capacity of crude oil distillation units alone, has appeared relatively stable in recent years, at about 18 million barrels per day of operable capacity. While the level is a reduction from the capacity of twenty years ago, the first refineries that were shut down as demand fell in the early 1980s were those that had little downstream processing capability. The small facilities, limited to simple distillation, became uneconomical as the need for downstream processing became a necessity. Additional refineries were shut down in the late 1980s and during the 1990s because of unprofitability and, at the same time, refiners improved the efficiency of crude oil distillation units that remained in service by debottlenecking to improve the flow and to match capacity among different units and by turning more and more to computer control of the overall processing.

Furthermore, following the need for environmentally benign products as well as commercial economics, refiners enhanced their upgrading (downstream processing) capacity. As a result, the capacity of the downstream units ceased to be the constraining factor on the amount of crude oil processed through the crude oil distillation system. In fact, crude oil input to refineries has continued to rise, and along with them refining capacity rose throughout much of the 1990s and refinery utilization reached record levels in the last half of the decade. In the United States, the Gulf Coast is the leader in refinery capacity and is the nation's leading supplier in refined products as in crude oil.

6.6.3 World Refining Capacity

While the Mideast is the largest producing region, the bulk of refining takes place in the United States, Europe or Asia because it was cheaper to move crude oil to the consuming areas rather move the products and the proximity to consuming markets made it easier to respond to weather-induced spikes in demand or to gauge seasonal shifts. Thus, the largest concentration of refining capacity is in the United States, accounting for about one-quarter of the crude oil distillation capacity worldwide. Also, the United States has by far the largest concentration of processing units (downstream capacity) to maximize the production of gasoline.

In 2007, the world refining capacity was 87,913,000 barrels per day, which is marginally lower than what is currently required although refinery throughput for 2007 was 75,545,000 barrels per day — an 86% usage rate (*BP*, 2008). Furthermore, world oil consumption for 2007 was 85,220,000 barrels per day, with the difference between consumption and refinery throughput being the amount of crude oil used directly as fuel by various industries. With another eight million to ten million barrels per day capacity coming on line in the next four to five years has the potential to retain an excess of refining capacity over production. However, the situation is dependent on the rate at which the product demand growth continues.

Constrained refining capacity has been frequently identified as one of the key reasons for the rise in crude oil prices over the last two years but refining capacity will begin to expand later this year and continue well past 2009 (http://zawya.com/printstory.cfm?storyid=v51n31-1TS05&l=133200080804). Most of the capacity will come on-stream in Asia and the Middle East and is driven by demand for middle distillates and the new refineries will have an upgrading capacity that is going to make it a bit easier to meet gasoil demand. Expanding global upgrading capacity promises to make a considerable difference in the market increasing the ability to upgrade heavy, high sulfur feedstock into distillate.

In addition to API gravity and sulfur content, the chemical composition and other natural characteristics of crude oil affect the cost of processing or restrict the suitability of crude oil for the manufacture of specific products and also dictates whether a crude oil feedstock can be used for the manufacture of specialty products, such as lubricating oils and/or of petrochemical feedstocks.

Refiners therefore strive to run the optimal mix of crudes through their refineries, depending on the refinery's equipment, the desired output mix, and the relative price of available crudes. In recent years, refiners have confronted two opposite forces: (1) consumer and government mandates that increasingly required light products of higher quality (the most difficult to produce) and, (2) crude oil supply that was increasingly moving towards the heavy oils having lower API gravity and higher sulfur content.

6.6.4 Refining and Refinery Economics

A petroleum refinery is an installation that manufactures finished petroleum products from crude oil, unfinished oils, natural gas liquids, other hydrocarbons, and alcohol. Refined petroleum products include but are not limited to gasoline, kerosene, distillate fuel oils (including No. 2 fuel oil), liquefied petroleum gas, asphalt, lubricating oil, diesel fuel, and residual fuel.

A typical large refinery costs billions of dollars to build and millions more to maintain and upgrade. It runs around the clock 365 days a year, employs between 1,000 and 2,000 people, and can occupy as much as one section (1 square mile, 640 acres) of land. Many refineries are of such a size that workers ride bicycles, or some other convenient form of transportation, to get from one station to another.

The complexity of refinery equipment varies from one refinery to the next. In general, the more sophisticated a refinery, the better its ability to upgrade crude oil into high-value products. However, whether simple or complex, all refineries perform three basic steps: separation, conversion, and treatment.

Thus, although the physical and chemical characteristics of crude oils differ, the core refining process is simple distillation (or separation) (Figure 6.1). Because crude oil is made up of a mixture

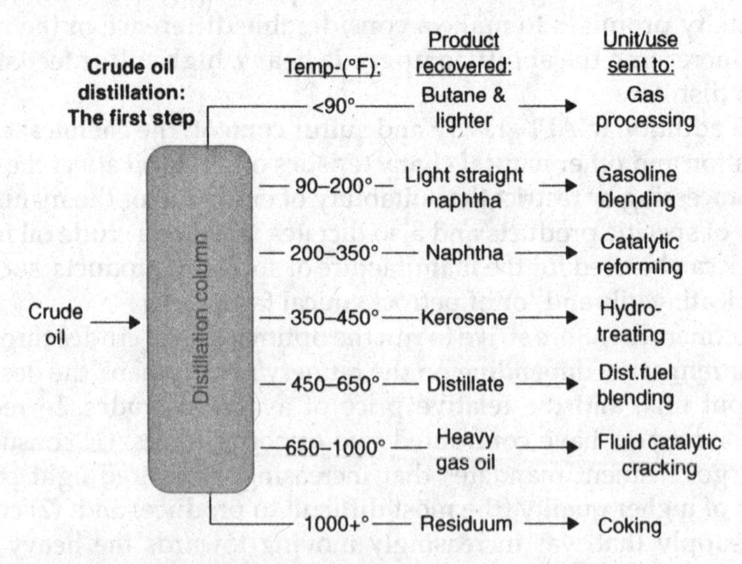

Figure 6.1 Schematic of crude oil distillation.

of hydrocarbons, this first and basic refining process is aimed at separating the crude oil into its fractions or broad boiling ranges of the various component hydrocarbons. Crude oil is heated and put into a still (distillation column) and different products boil off and can be recovered at different temperatures. The lighter products, including liquid petroleum gases (LPG), naphtha, and so-called straight run gasoline (i.e. the gasoline fraction distilled directly from the crude oil without further processing), are recovered at the lowest temperatures. Middle distillates; such as jet fuel, kerosene, home heating oil, and diesel fuel distill next. Finally, the highest molecular weight fractions (residual fuel oil and residuum) are recovered. Additional processing follows crude distillation and involves the use of a variety of highly complex units designed for very different upgrading processes to produce the most desirable lower molecular weight products.

At this point, the lower boiling fractions from the distillation tower are sent to treating units while the higher boiling fractions are converted into streams (intermediate components) that eventually become finished products, such as gasoline.

The most widely used conversion method uses high temperature and pressure to convert (crack) higher molecular weight (high-boiling) hydrocarbons into lower molecular weight (low-boiling) products. Fluid catalytic cracking (cat cracking), is the basic gasoline-making process. Using heat of approximately 500°C, 932°F, low pressure, and a catalyst (a substance that accelerates chemical reactions), the cat cracker can convert most relatively heavy fractions into lower molecular weight gasoline constituents. Hydrocracking applies the same principles but uses a different catalyst, slightly lower temperatures of approximately 450°C, 842°F, much greater pressure, and hydrogen to obtain conversion to the desired products.

Some refineries employ coking units, which use heat of approximately 500°C, 932°F to convert the distillation residuum into lower-boiling products and coke, which may be used as an industrial fuel.

Cracking and coking are not the only forms of conversion and other refinery processes, instead of splitting molecules, rearrange them to add value. Alkylation, for example, makes gasoline components by combining some of the gaseous byproducts of cracking. Reforming uses heat, moderate pressure and catalysts to turn naphtha, a low-boiling, relatively low-value fraction, into high-octane gasoline components. To make gasoline, refineries carefully combine a variety of streams from the processing units to produce the blended product that is sold as gasoline. The finished gasoline products may contain more than 200 individual hydrocarbons and additives so that the product meets the designated specifications.

The quality of the crude oil dictates the level of processing and reprocessing necessary to achieve the optimal mix of product output. Hence, price and price differentials between crude oils also reflect the relative ease of refining. A premium crude oil, such as West Texas Intermediate (WTI), the U.S. benchmark crude oil (Chapter 4), has a relatively high natural yield of desirable naphtha and straight-run gasoline. nother premium crude oil, Nigeria's Bonny Light crude oil, has a high natural yield of middle distillates. By contrast, almost half of the simple distillation yield from Arabian Light crude oil, the historical benchmark crude, is a heavy residue, or "residuum," that must be reprocessed or sold at a discount to crude oil. Even West Texas Intermediate and Nigeria's Bonny Light crude oil have a yield of about one-third residuum after the simple distillation process.

In recent years, growth in the demand for petroleum products has led to an improvement in capacity utilization, increasing operating efficiency and reducing costs per unit of output. Refinery capacity is based on the designed size of the crude distillation unit(s) of a refinery — often referred to as nameplate capacity). Occasionally, through upgrades or de-bottlenecking procedures, refineries can process more crude than the nameplate size of the distillation unit would indicate. In such cases, a refinery is able to achieve a utilization rate greater than 100% for short periods of time.

Finally, not all investment decisions are driven by refinery economics; refiners also make investment decisions because of voluntary actions or legislative and regulatory requirements. In recent years, governments and industry have directed considerable effort towards reducing the environmental impact of burning fossil fuels. Many of the initiatives have been aimed at providing cleaner fuels. Petroleum refining is a very complicated and capital-intensive industry. New environmental regulations require the industry to make additional investments to meet the more stringent standards.

The ultimate operating variable in the economics of refining is the price of crude oil. Heavy high-sulfur crude oil can cost up to one-third less than lighter, low-sulfur crude oil but because high-sulfur crude oil requires more processing, refineries that buy these cheap crude oils usually higher fixed expenses for equipment and labor. In fact, processing high-sulfur crude oil requires greater expenditures for energy and energy may account for up to 50% of the cost of running a refinery. Another variable is the location of the refinery insofar as the closer a refinery is to the crude oil source and a high demand market, the lower transportation costs for incoming feedstock and outgoing products.

6.7 Outlook

Crude oil is the lifeblood of the economy of the United States and, for over a century, has played a dominant role in the economy. Oil currently accounts for close to 40%of total energy consumption in the US. However, while the U.S. economy grew significantly at 63%, after 1985, oil consumption grew much more slowly at 25%. Thus, our economy is far less dependent on oil than it was in 1973, the year of the Arab oil embargo.

Nonetheless, as the United States emerges from the current recession and the economy grows over the next few decades, the demand for oil will also grow. According to the Energy Information Administration (EIA), demand for oil in the United States is expected to rise from an annual average of 19.7 million barrels per day to more than 26 million bpd in 2020. At the same time, forecasts indicate that domestic oil production in the United States will likely see little, if any, growth; it is more likely to see reduction in view of the poor judgment by the United States Congress in not seeking alternate sources of energy. Therefore, the United States will increasingly rely on foreign sources to meet its oil needs. Currently, more than 65% of the crude oil and crude oil products used in the United States is imported from foreign sources. And, over the next two decades, imports of petroleum and petroleum products are expected to increase by more than 6 million barrels per day as oil consumption rises.

The historical record shows substantial variation in world oil prices, and there is the likelihood of even more uncertainty about future prices when longer time periods are examined. It is possible to consider three price cases to illustrate the uncertainty of prospects for future world oil resources (Annual Energy Outlook, 2008). In the reference case, world oil prices increase moderately from current levels (ca. $39 per barrel) to about $57 per barrel in 2016, start rising again as production in non-OPEC regions peaks, and continue rising to $70 per barrel in 2030 (all prices in 2006 dollars).

The low and high price cases reflect a wide band of potential world oil price paths, ranging from $42 to $119 per barrel in 2030, but they do not bound the set of all possible future outcomes. The high and low oil price cases are predicated on assumptions about access to and costs of non-OPEC oil, OPEC supply decisions, and the supply potential of unconventional liquids.

The world counts on the Organization of Petroleum Exporting Countries (OPEC) for more than 40% of total daily production of crude oil. However, there is some doubt about the oil reserve estimates of the OPEC member countries. Saudi Arabia, for example, has claimed crude reserves of 260 billion barrels for the past decade despite having pumped between nine and 10.5 million bpd throughout that period. In fact, the Paris-based International Energy Agency (IEA) predicts that the world can continue to increase oil production for the next 25 years, but at a significant investment that may be in excess of $600 billion US per year for the entire period. Added to this are fears for political stability in several oil-producing states.

In terms of actual reserves numbers within the OPEC membership, Saudi Arabia possesses the largest resource base in the world, with claims of 268 billion barrels of proven reserves. Other OPEC countries also show promise of producing higher quantities of oil.

However, assuming no serious political crises in key producing countries or an unexpected shortfall in investment, global oil production capacity will continue to grow strongly toward 102.4 million barrels per day by 2010 from the current level of approximately 87 million barrels per day. This expansion will be fairly evenly split between OPEC and non-OPEC countries: 8.5 million barrels per day and 6.7 million barrels per day, respectively. The expansion continues to 2015, but OPEC shows a greater increase: a net gain of 12.2 million barrels per day (relative to 2005) versus 8.2 million barrels per day for non-OPEC. At the regional level, the United States and North Sea show decreases through 2015, while Canada, West and North Africa, Latin America, and the Caspian, and the Middle East continue their current trend of strong expansion past 2010 and through 2015. Southeast Asia shows some modest growth, but declines after 2010. At the same time, Russian capacity growth slows.

By 2015, there could be a change in the geographic focus of the sources of liquids supply. The proportion of liquids capacity from the top 15 countries will rise from less than the current 60% to 65% in 2015. While nearly every OPEC country, except Indonesia, shows potential for a significant increase to 2015, the sources of expansion in non-OPEC countries are more limited, with Russia, the Caspian nations, Brazil, Angola, and Canada leading the way. There may also be the emergence of some new sources of liquids capacity both in the deep water, such as offshore Mauritania, and onshore in Sudan. In addition, mature areas, such as Malaysia, are reemerging and a new play is being successfully developed in a previously unexplored deep-water area of offshore Sabah, in northwest Malaysia. However, this shift in emphasis may prove to be to more politically and operationally challenging countries, which increases the levels of risk and supply anxiety in some consumer countries.

Analysis of the composition of new capacity shows that in the medium term, there will be increasing proportions of light and heavy oils and a reduction in the proportion of medium grade crude. However, capacity additions to 2010 are predominately light crude oil (8 million barrels per day), medium crude oil (5 million barrels per day), and heavy crude oil (3 million barrels per day) with a continuing rapid expansion of deep-water production capacity to over 9 million barrels per day by 2010.

Production capacity of heavy oil and tar sand bitumen from Canada and Venezuela will expand to approximately 5 million barrels per day in 2015. The Canadian projects are moving forward at an accelerating pace and expansion from 1 million barrels per day of synthetic crude oil (from tar sand bitumen) to more than 3 million barrels per day by 2015 is anticipated, with approximately half being mined and the remainder in situ. In Venezuela the four main Orinoco projects are on-stream (totaling 650,000 barrels per day) and with debottlenecking could reach 700,000 barrels per day by 2010.

Between 2010 and 2015, there is considerable potential to expand total condensate plus natural gas liquids (NGLs) capacity to 22 million barrels per day. Notable condensate expansions will occur in Qatar as the liquefied natural gas (LNG) business expands and more gas is produced for pipeline exports and gas-to-liquids conversion. One of the largest expansions of condensate capacity will occur in Norway.

Until recently the gas-to-liquids (GTL) business had contributed only a small proportion of production at less than 200,000 barrels per day in 2008, but there are a number of projects under way and planned that are expected to boost production capacity to 480,000 barrels per day by 2010 and 1 million barrels per day in 2015. This is a lower buildup than might be anticipated by summing the reports of current activity, but operators may not commit to new gas-to-liquid projects until there is some certainty that the oil price will remain high enough for profitability.

The question of a worldwide peak in oil production continues to stimulate debate, although there is little evidence for a peak in worldwide oil production before 2020. It is true that total annual global production has not been replaced by exploration success in recent years, but production has been more than replaced by exploration plus field reserve upgrades. Although oil is a finite resource, there is still no exact estimate of total reserves; meanwhile global resources should continue to expand. Many basins, even those producing significant volumes of oil, remain underexplored.

One of the reasons that there is so much debate over the whether the peak of oil production is imminent or not is that different observers rely upon different data sets. The most visible data are those published by exploration and production companies through annual reports and the most extensive collection of such reports are the filings under United States Securities and Exchange Commission (SEC). However, these data may be overly conservative as evidenced by the extent to which upward reserve revisions outweigh downward revisions.

In fact, it is entirely possible that large portions of discovered fields maybe excluded from disclosure until later in their producing lives and that only a small portion of the overall picture is revealed by these disclosures. As often happens with regulatory systems that have been in place for three decades, it requires modernization to take into account almost revolutionary changes in technologies, and transformations in terms of market structure and geography.

If the SEC were to adopt the definitions and guidelines used by the Society of Petroleum Engineers (SPE), this would lead to the creation of a globally consistent data set that covered the vast majority of the world's oil and gas reserves. As the very definition of what is oil begins to change with the addition of non-traditional resources such as synthetic crude oil (Syncrude) and gas-to-liquids and even liquid fuels made from coal, a reliable dataset will be lead to a better understanding of when the world may face an undulating plateau of global oil production.

Finally, in consideration of the above effects, forecasting crude oil economic activity has received considerable attention over the past several decades. An increasing number of statistical methods, which frequently differ in structure, have been developed in order to predict the evolution of various macroeconomic time series and commodity prices have been the focus of various studies (Roche 1995; Labys 1999; Morana 2001; Lanza *et al*, 2005). However, whatever techniques are employed to forecast future spot prices, data for short-term horizons offer some indications of prices whereas for longer-time horizons the data are less accurate (Fernández, 2006). In fact forecasting crude oil prices is an imprecise art. One only need consider the price fluctuations during 2008 to realize that no one predicted any such fluctuations and the price forecasting was inaccurate not only because of the methods employed, but because of many unknowns that are likely to arise at any time and for any reason.

6.8 References

Annual Energy Outlook. 2008. Annual Energy Outlook 2008 with Projections to 2030. Report No. DOE/EIA-0383(2008). Energy Information Administration, Department of Energy, Washington, DC. June.

Balke, N.S., Brown, S.P.A., and Yücel, M.K. 1998. Economic Review First Quarter 1998. Federal Reserve Bank of Dallas, Dallas Texas.

BP. 2008. BP Statistical Review of World Energy. Www.bp.Com/statisticalreview.

Fernández, V. 2006. Forecasting Crude Oil and Natural Gas Spot Prices by Classification Methods. Documentos De Trabajo with Number 229. Centro De Economía Aplicada, Universidad De Chile, Santiago, Chile. http://ideas.repec.org/p/edj/ceauch/229.html

Gonzalez R. 2006. Oil Industry Investment. Petroleum Technology Quarterly 8(4): 3.

IEA. 2008. Key World Energy Statistics 2008, International Energy Agency, Paris, France.

ITF. 2008. Interim Report on Crude Oil. Interagency Task Force on Commodity Market s , Washington DC. July.

Labys, W. 1999. Modeling Mineral and Energy Markets. Kluwer, New York, USA.

Lanza, A., Manera, M., and Giovannini, M. 2005. Modeling and Forecasting Co-integrated Relationships among Heavy Oil and Product Prices." Energy Economics, 27(6):831–848.

Laufenberg, D.E. 2007. Higher oil prices are not always shocking. Report No. 231483 A (1/07). Ameriprise Financial Inc., Minneapolis Minnesota.

Morana, C. 2001. A Semiparametric Approach to Short-Term Oil Price Forecasting. Energy Economics 23(3), 325–338.
Peterson, D.J., and Mahnovski, S. 2003. New Forces at Work in Refining: Industry Views of Critical Business and Operations Trends. Prepared for the National Energy Technology Laboratory United States Department of Energy. Rand Corporation, Santa Monica, California.
Roche, J. 1995. Forecasting Commodity Markets. Probus Publishing Company, London, England.
Swain, E.J. 1991. Oil & Gas Journal. 89(36): 59.
Swain, E. J. 1993. Oil & Gas Journal. 91(9): 62.
Swain, E. J. 1997. Oil & Gas Journal. 95(45): 79.
Swain, E.J. 2000. Oil & Gas Journal. March 13.

7

The Crude Oil Market

Russia, Saudi Arabia, and the US are the largest producers of crude oil in 2008. The greatest concentration of production (both oil and gas) is in the Middle East. Furthermore, since the discovery of crude oil in the late 19th century, the world has consumed approximately 1.0 trillion (1×10^{12}) barrels of crude oil. There are sufficient quantities (1.0 trillion barrels) of proven conventional crude oil reserves (determined by drilling) to meet today's global demand for another 25 years. There are expected to be another 2 to 3 trillion barrels of unproven conventional crude oil reserves, such as the heavy Venezuelan and Canadian crudes, and shale oil (Chapter 9), which are together much more costly to produce than conventional oil. So, although the world may not be approaching peak oil (Chapter 8), it is much more likely to be approaching peak cheap oil.

The world is depleting the reservoirs of light sweet crude oil, which can be produced at both low costs and at low business risk. Much of the conventional crude resource oil is located in politically unstable countries (Russia, Middle East and West Africa), which adds a price premium to the cost of production. Furthermore, the crude oil market in the United States is the largest oil market in the world and accounts for almost 24% of world oil demand (BP, 2008). Gasoline accounts for about 50% of total US oil demand at approximately 9 million barrels per day. The US oil market is very important not only because of its size, but also because it is a center of oil price formation led by oil price determination in the NYMEX (New York Mercantile Exchange) oil futures market.

In spite of predictions of gloom-and-doom for remaining petroleum resources, gasoline consumption in the United States has been growing steadily in recent years, backed by such factors as strong economic growth and the popularity of less fuel efficient vehicles like sports utility vehicles (SUV). The recent turndown in the economic climate has curbed this demand somewhat but demand is expected to recover and increases in the short term.

However, demand for crude oil in the United States continued to grow steadily led by gasoline demand growth, and bottlenecks emerged in the oil product supply chain. First, domestic oil refineries in the United States are operating at full capacity with the refining utilization rate exceeding 90% on annual average. Second, surplus production capacity for petroleum products in the United States is very limited. Finally, oil inventories in the private sector have remained low because of oil industry efforts to reduce costs.

In terms of some surplus refining capacity in the United States (Chapter 6) and the issues regarding oil product specification, the problems continue in the United States market. The downstream oil market plays an important role in stabilizing supply and demand in terms of the share of the world market, because (1) the United States is the world's largest oil consumer and the oil price determination in US (NYMEX) influences the world oil prices, and (2) the oil product supply-demand balance in the United States remains tight as there is little or no spare domestic refining capacity.

In addition, investor constraints also exist for refinery investments, although this may differ in existence and degree by economy. Some of these constraints are:

1. Low return on investments including low refining margin, huge upfront investment costs, high risk due to instability and uncertainty of crude oil supply
2. Site acquisition and permitting problems for refinery construction
3. Stringent environmental standards and the changing fuel specifications.

There is some potential for refinery expansion and/or refinery reactivation based on current and future oil demand growth. But, there is uncertainty that (1) the planned refinery investment is realized, and (2) the planned refinery expansion/upgrading are sufficient to meet the growing and changing oil demand.

These are only some of the factors that influence crude oil and product, particularly gasoline, pricing and there is a growing need to understand the issues related to pricing in the crude oil and gasoline markets. While many factors exist, the predominant factors are presented below.

7.1 The Crude Oil Market

The crude oil market is complex and depends on several facets, each of which can move or disturb the market either in favor of higher prices or in favor of lower prices. The crude oil market is made up of producing-exporting nations (e.g., the OPEC member nations) and the consuming-importing nations.

It is no secret that the three largest crude oil consuming regions — North America, Europe, and Asia-Pacific — are all oil importers, while all of the other regions are exporters. The Middle East exports significantly more oil than any other region, despite the strong growth in production in other areas in recent years and the global dependence on the Middle East region for oil makes the geo-political importance of the Middle East readily understandable.

On the other hand, there is less variation among the importing regions. The Asia-Pacific economies have propelled the region into the top place with import growth double that of the import growth of any other region. Even though the United States is the largest individual importer, as a region North America as a region ranks third; Canada and Mexico are two of the top three crude oil suppliers to the United States and the imports of oil from these countries into the United States offsets imports from these neighbors in the regional calculation. This has kept North America's import artificially low — a story not always shared or publicized by agencies that report monthly global imports and exports of crude oil.

Crude oil is the highest traded commodity on the international market, whether it is measured by volume, by value, or by the carrying capacity needed to move it.

Generally, crude oil and petroleum products flow to the markets that provide the highest value to the supplier. Everything else being equal, oil moves to the nearest market first, because that has the lowest transportation cost and therefore provides the supplier with the highest net revenue, or netback. If this market cannot absorb all the oil, the balance moves to the next closest one thence to the next, incurring progressively higher transportation costs, until all the oil is placed.

Because of the political instability of the Middle East, policy makers in the United States have viewed this increased dependence on the Western Hemisphere as the major supplier of crude oil to the United States, and the decreased dependence on the Middle East crude oil, as a welcome development. In fact, the recent growth in United States dependence on its Western Hemisphere neighbors is considered by some to be an illustration of this nearer-is-better syndrome because the Western Hemisphere sources now supply over half the United States import volume, much of it on voyages of less than a week. Another quarter comes from elsewhere in the Atlantic Basin (countries on both sides of the Atlantic Ocean) and takes just two to three weeks to reach the United States. Another view is that the importers in the United States are also seeking sources of oil that are, for the most part, controlled by relatively stable governments.

As a result of the shift by the United States toward sources of crude oil in the Western Hemisphere, Saudi Arabia is the only significant Middle East supplier left. Although the dependence of the United States on the long-haul Middle East has fallen sharply, this has not made prices in the United States vulnerable to a disruption in Middle East supplies.

Mexico and Venezuela have consciously helped the trend toward short-haul shipments.They pro-actively took the strategic decision to make as large and as profitable a market as possible for poor quality crudes, since their reserves are unusually biased toward those hard-to-place grades. Both countries began with refineries that had traditionally run their own crudes, and then with refineries that might be upgraded to handle heavier crude oils. This has turned low API gravity high sulfur crude oil into the preferred crude oil at these sites, significantly increasing the crude oil self-sufficiency of the Western Hemisphere. Nevertheless, the political stability of both governments is always open to speculation. However, in practice, the direction of crude oil trade does not always follow the nearest first pattern. Refinery configuration, product demand, product quality specifications and politics can all change the rankings.

In fact, different markets frequently place different values on particular grades of crude oil. For example, crude oil, which is more amenable to the production of low sulfur diesel, is worth more in the United States than it is in countries the maximum allowable sulfur is much higher. From another aspect, low sulfur crude oil is relatively more expensive in countries where the crude oil allows a refiner to meet tighter sulfur limits in the region without investing in refinery upgrades. These differences in the value of crude oil quality can be sufficient to overcome transportation costs. However, government tariffs on crude oil can also influence the quality of crude oil imported into some countries and may even negate some of the benefits derived from quality.

The crude oil market can also be dominated by refinery placement (Chapter 6) insofar as the site of refinery placement (i.e. closer to the consumer market rather than close to the wellhead) takes maximum advantage of the economies of scale of large ships, especially as local quality specifications are increasingly fragmenting the product market. The placement of a refinery close to the consumer market maximizes the ability of the refinery to tailor the product output to the market by accommodating any short-term surges such as those caused by weather, equipment outages, and other events. In addition, this policy also guards against the very real risk that governments will impose selective import tariffs and/or restrictions to protect the domestic refining sector.

However, there are some refineries that are exceptions to this generality, having been developed to serve particular export markets. These export refining centers, such as refineries in Singapore, the Caribbean, and the Middle East, give rise to some regular inter-regional product moves, but they are the exception. The inter-regional products trade is largely a temporary function set in place to balance market demand as might occur when there is a high demand for heating oil due to colder than predicted winters.

Transport to the market has already been mentioned but a further comment is warranted to note that there are two modes of transportation for inter-regional trade: tankers and pipelines (Chapter 3). Tankers have made global (intercontinental) transport of oil possible, and they are low cost, efficient, and extremely flexible, but on the other hand, a pipeline is the method of choice for transcontinental movement of crude oil.

With respect to tankers, each route usually has one size that is economic appropriate and is based on the following factors:

1. Voyage length
2. Port constraints
3. Canal constraints
4. Volume.

For example, crude oil exports from the Middle East (involving high volumes and long distances) are carried mainly by very large crude carriers (VLCCs), which typically carry more than two million barrels of oil on every voyage. The economies of scale of the very large crude carriers outweigh the constraints imposed by size differences and a long voyage is often cheaper than a short voyage when the cost is compared on a on a per barrel basis. The very large crude carriers are too large for all the ports in the United States except the Louisiana Offshore Oil Port (LOOP) and must have some or all of their cargo transferred to smaller vessels either at sea (lightering) or at an offshore port (transshipment). In contrast, ships out of the Caribbean and South America are routinely smaller and can enter ports in the United States without the need for offshore unloading of the crude oil.

On the other hand, pipelines are critical for transportation of landlocked crude oil and also complement tankers at certain key locations by relieving bottlenecks or providing shortcuts. In fact, the only inter-regional trade that currently relies solely on pipelines is crude from Russia to Europe. Export pipelines are also needed for production from the Caspian Sea region but the greatest negative for pipelines crossing national boundaries is the vulnerability of the pipeline to political and terrorist acts. However, pipelines are the primary option for transcontinental transportation because they are at least an order of magnitude cheaper than any alternative such as rail, barge, or road. In such cases, pipeline vulnerability is usually small or non-existent within national borders (such as the United States and Canada). In fact, the development of large diameter pipelines during World War II allowed the development of the vast pipeline network in North America that moves crude oil and product within Canada, from Canada into the United States, and within the United States. Pipelines are also an important oil transport mode in mainland Europe, although the system is much smaller, matching the shorter distances.

As already noted, the United States is the major importer of crude oil and accounts for approximately 25% of total world imports of crude oil. Yet its import dependency, the percentage of demand met by imports, is significantly lower, at about 50% than that of its international partners. Industrialized countries such as Japan and Germany have import dependency levels of 90 to 100%.

Canada is the one country that delivers oil to the United States by pipeline because the majority of crude oil in Canada is land-locked and rely almost exclusively on pipelines from Western Canada that tie into the transcontinental pipeline network to reach their main export markets, which lie all across the Northern Tier of the United States.

The trade among regions of the United States is focused on the eastern half of the country. The Gulf Coast is by far the largest supplier, accounting for more than 80% of the flow in the Petroleum Administration for Defense Districts (PADD). In contrast, the Rockies and the West Coast are isolated, in petroleum logistics terms, from the rest of the country. The relatively easy flow of petroleum from the Gulf Coast to the Midwest and the East Coast mean that incremental supply is more readily available to those markets in the event of a demand surge or supply drop.

7.2 Global Oil Consumption

Determining oil consumption presents an interesting problem. The size and complexity of the market and the number of consumers and suppliers make data collection a daunting task and a variety of approaches are necessary in order to measure (or, at best, estimate) consumption of crude oil. In addition, because of the variations in climate and topography within the United States (also including Alaska and Hawaii), regional markets exhibit different oil consumption patterns. Population and regional economic activity are two important determinants, but the traditional availability of alternative fuels, petroleum transportation, geography, and a host of other factors are also important.

The most convenient method and perhaps the most accurate is to use the following criteria:

1. Refinery input
2. Refinery production
3. Product imports
4. Changes in inventory.

Using the data, it is possible to arrive at general conclusions about the market demand and use of petroleum and petroleum products. While the data may not be sufficient to the nearest ten barrels of oil, the result will be a rage that can be used to differentiate between changes in market supply and demand over a given period of time.

The market demand for crude oil is derived from the demand for the finished products and, consequently, the industrialized countries are the largest consumers of oil and the countries of the Organization for Economic Cooperation and Development (OECD) account for almost more than 60% of worldwide daily oil consumption. It is not surprising that the developed economies use oil much more intensively than the developing economies with oil consumption in the United States being on the order of 2.5 gallons per capita per day. Of course, average often bare no relationship to reality and the oil consumption among the high salaried multi-vehicle families will be as much as double the average with an accompanying reduction in oil consumption by lower-income single vehicle families. No-vehicle families, being users of public transportation systems, will have the lowest daily per capita consumption of petroleum. Generally, the United States uses more crude oil for the production of transportation fuels than for the production of fuels for heat and power. However, the precise amount of crude oil used for the various purposes varies from year-to-year, depending on weather and economic activity.

In the years since the Arab Oil Embargo of 1973/74, transportation has been a major component of market demand. The demand for non-transportation uses of oil, such as the production of heating fuels, has also continued but not to the same extent as gasoline. Overall, the transportation sector accounted for 49% of world oil consumption in 2005 at 38.3 million barrels of oil equivalent (mboe) per day, up from a share of only one third in 1971

(Figure 2.1). This share is set to continue to rise, reaching 52% by 2030. The growing importance of the transportation sector to oil demand is unsurprising, given the limited fuel switching possibilities and the expected continued growth in people's mobility (BP, 2008).

The market for distillate fuel oil use ranks second behind gasoline. Unlike gasoline, which is used almost exclusively in the transportation sector, distillate fuel oil is used more widely for the following purposes:

1. Home heating fuel
2. Industrial power
3. Electric generation
4. Diesel-fueled vehicles.

Diesel fuel used in vehicles on the highway for trucks, buses, and passenger cars must be low or no sulfur, while distillate fuel oil used off the highway for vessels, railroads, farm equipment, industrial machinery, electric generation, or space heating is not subject to the same low sulfur highway standard, but as a matter of course contains only a small amount of sulfur. The Unites States Department of the Treasury also requires that the non-highway product must be dyed to distinguish it from the taxable on-highway diesel. These requirements also limit distribution flexibility for distillate fuels, requiring segregated storage and transportation, and preventing one product from easing shortages of the other.

The market for jet fuel is the third-highest product in demand and, like gasoline, is largely confined to use in the transportation sector. The military formerly utilized a different, naphtha-based product for aircraft, instead of the commercial, kerosene-based product, but in recent years, has converted to kerosene-based jet fuel.

Residual fuel oil is the heavy fuel used in boilers for power generation and to propel tankers and other large vessels, but the market has been eroded by a variety of factors, including price competition with newly available natural gas and environmental restrictions. Residual fuel oil's use for apartment building space heating is now confined largely to older buildings in New York City.

Market prices of crude oil have varied over the past 25 years (Table 7.1; Figure 7.1) and, like those of other goods and services, reflect both the product's underlying cost as well as market

Table 7.1 Crude oil price highs and lows over the past 25 years.

1984	28
1988	15
1990	23
1994	16
1996	20
1998	12
2000	27
2002	23
2008	147
2009	41

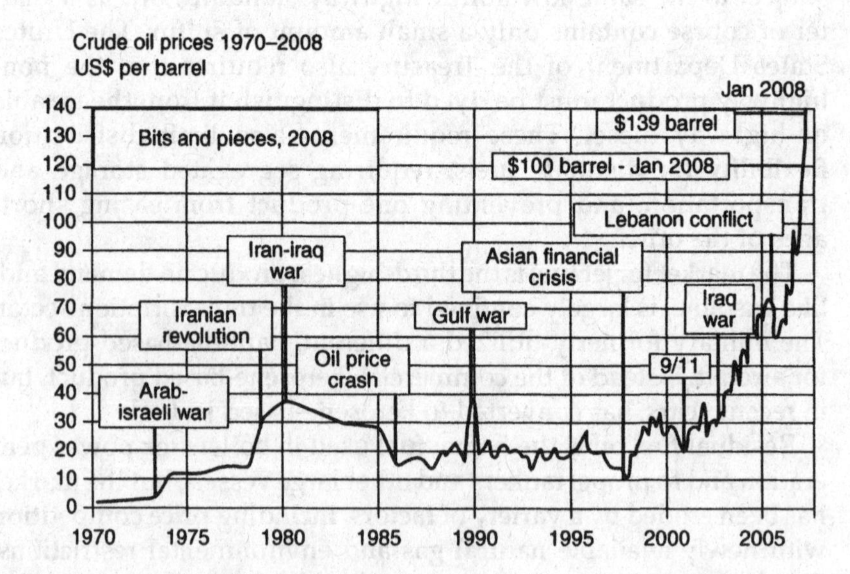

Figure 7.1 Crude oil price variations since 1970.

conditions at all stages of production and distribution. For example, he pre-tax price of gasoline or any other refined oil product reflects the following:

1. The price of crude oil
2. Transportation from the producing field to the refinery
3. Refining
4. Transportation from the refinery to the market
5. Transportation, storage and distribution between the market distribution center and the retail outlet or consumer
6. Market conditions at each stage along the way, and in the local market.

Oil markets are essentially a global auction and the price of crude oil, the raw material from which petroleum products are made, is established by the supply and demand conditions in the global market overall, and more particularly, in the main refining centers. In fact, crude oil prices are a result of thousands of transactions taking place simultaneously around the world, at all levels of the distribution chain from crude oil producer to individual consumer.

Furthermore, there are several different types of transactions that are common in oil markets. Contract arrangements in the oil market in fact cover most oil that changes hands. Oil is also sold in spot transactions (cargo-by-cargo, transaction-by-transaction arrangements). In addition, crude oil is traded as a commodity in futures markets, which are a mechanism designed to distribute risk among participants on different sides, such as buyers versus sellers, or with different expectations of the market, but not generally to supply physical volumes of oil. Both spot markets and futures markets provide critical price information for contract markets.

Prices in spot markets are considered to be a signal of the supply/demand balance. Rising prices generally indicate that more supply is needed, and falling prices indicate that there is too much supply for the prevailing demand level. Furthermore, while most oil flows under contract, the price varies with spot markets. Futures markets also provide information about the physical supply/demand balance, as well as the market's expectations.

Seasonal swings are also an important underlying influence in the supply/demand balance, and hence in price fluctuations.

Other things being equal, crude oil markets would tend to be stronger in the fourth quarter (the high demand quarter on a global basis, where demand is boosted both by cold weather and by stock building) and weaker in the late winter as global demand falls with warmer weather. As a practical matter, crude oil prices reflect more than seasonal factors and are, in fact, subject to a host of other influences. Likewise, product prices tend to be highest, relative to crude oil, as they move into their high demand season – late spring/early summer for gasoline, late autumn for heating oil. The seasonal pattern in actual product prices, again, may be less obvious, because so many other factors are at work.

Price change patterns can vary between regions, depending on the prevailing supply/demand conditions in the regional market, especially in the short-term. Both geography and the unique quality of the gasoline contribute to the volatility of gasoline prices. If sources for additional supply are limited, any unusual increase in demand or reduction in supply gets a large price response in the market. The price response, and the differences in regional price movements, is critical to the way the oil market redistributes products to re-balance after an upheaval. A price increase in one area calls forward additional supplies, which might come from other markets in the United States, or from incremental imports or they may also be augmented by increased output from refineries.

Ultimately, oil prices can only be as high as the market will bear, as was evident during the late summer of 2008. They may be higher in areas with higher disposable income, where real estate values, wages and other measures of economic activity indicate that the market is more robust. However, if they rise higher than the market will bear, consumers will seek substitutes or downsize their cars and make other adjustments that reduce their consumption. If the local area offers unusually high profits, competitors will quickly enter the market, finally pushing prices down.

Crude oil prices are the most important determinant of petroleum product prices, and often the most important factor in price changes as well. Crude oil prices reflect an overall market balance — when crude oil prices are low, reflecting an oversupply, product prices will also be low; when crude oil prices are high, reflecting undersupply or high demand, product prices will also be high. When the price of crude oil moves up or down on a sustained basis, the change will be reflected in product markets, all other things being equal.

7.3 Refining and The Markets

The petroleum refining industry converts crude oil into a wide slate of products, which include liquefied petroleum gas, naphtha, gasoline, kerosene, aviation fuel, diesel fuel, fuel oils, lubricating oils, wax, asphalt, and feedstocks for the petrochemical industry (Speight, 2007, 2011).

Petroleum refining can be traced back over 5000 years to the time of the Sumerians when asphalt materials and petroleum-type products were isolated not only from areas where natural seepage occurred but also by distillation (Hsu and Robinson, 2006; Speight 2007, 2011). Petroleum refining is a recent technology and many innovations evolved during the 20th century. Furthermore, as feedstocks change in character, these innovations will need to continue until petroleum refining as is currently known becomes applied to, and adapted to, other sources of energy (Speight, 2008).

Refinery activities start with the receipt of crude oil for storage at the refinery, followed by dewatering and desalting, and include all of the subsequent handling and refining operations. Refining terminates with storage preparatory to shipping the refined products from the refinery. The industry employs a wide variety of processes, and the composition of the crude oil feedstock and the chosen slate of petroleum products determine the processes selected from refining, or processing flow scheme. The selection and arrangement of these processes will vary among refineries, and few, if any, employ all of these processes.

A refinery is a group of integrated manufacturing plants that vary in number with the variety of products produced (Hsu and Robinson, 2006; Speight, 2007, 2011) and are selected to give a balanced production of saleable products in amounts that are in accord with the demand for each. To prevent the accumulation of non-saleable products, the refinery must be flexible and be able to change operations as needed. The complexity of petroleum is emphasized insofar as the actual amounts of the products vary significantly from one crude oil to another (Ancheyta and Speight, 2007; Speight, 2007, 2011).

The refining industry has been the subject of the four major forces that affect most industries and which have hastened the development of new petroleum refining processes:

1. The demand for products such as gasoline, diesel, fuel oil, and jet fuel
2. Feedstock supply, specifically the changing quality of crude oil and geopolitics between different countries and the emergence of alternate feed supplies such as bitumen from tar sand, natural gas, and coal
3. Environmental regulations that include more stringent regulations in relation to sulfur in gasoline and diesel, and
4. Technology development such as new catalysts and processes.

Petroleum refineries were originally designed and operated to run within a narrow range of crude oil feedstock and to produce a relatively fixed slate of petroleum products. Since the 1970s, refiners had to increase their flexibility in order to adapt to a more volatile environment. Refiners, to increase their flexibility within existing refineries, may use several possible paths. Examples of these paths are change in the severity of operating rules of some process units by varying the range of inputs used, thus achieving a slight change in output. Alternatively, refiners can install new processes and this alternate scenario offers the greatest flexibility, but is limited by the constraint of strict complementarily of the new units with the rest of the existing plant, and involves a higher risk than the previous ones. It is not surprising that many refiners decide to modify existing processes.

The means by which a modern refinery operates depends not only on the nature of the petroleum feedstock, or, nowadays, the different crude oils that make up the blended feedstock, but also on its configuration (i.e., the number of types of the processes that are employed to produce the desired product slate). This is strongly influenced by the specific demands of a market. Therefore, refineries need to be constantly adapted and upgraded to remain viable and responsive to ever changing patterns of crude supply and product market demands. As a result, refineries have been introducing increasingly complex in order to produce higher yields of, for example, hydrocarbon fuels, from higher boiling fractions and residua.

Whatever the choice, refinery practice continues to evolve and new processes are installed in live with older modified process. The purpose of this chapter is to present to the reader a general overview of refining that, when taken into the context of the following chapters, will show some of the differences that occurring in refineries.

The configuration of any refinery-type may vary from refinery to refinery. Some refineries may be more oriented toward the production of gasoline (large reforming and/or catalytic cracking), whereas the configuration of other refineries may be more oriented towards the production of middle distillates, such as jet fuel and gas oil.

A modern refinery is a highly complex and integrated system separating and transforming crude oil into a wide variety of products, including transportation fuels, residual fuel oils, lubricants, and many other products. The simplest refinery configuration is the topping refinery, which is designed to prepare feedstocks for petrochemical manufacture or for production of industrial fuels in remote oil-production areas.

The topping refinery consists of tankage, a distillation unit, recovery facilities for gases and light hydrocarbons, and the necessary utility systems of steam, power, and water-treatment plants. Topping refineries produce large quantities of unfinished oils and are highly dependent on local markets, but the addition of hydrotreating and reforming units to this basic configuration results in a more flexible hydroskimming refinery, which can also produce desulfurized distillate fuels and high-octane gasoline. These refineries may produce up to half of their output as residual fuel oil, and they face increasing market loss as the demand for low sulfur (even no sulfur) high sulfur fuel oil increases (Ansheyta and Speight, 2007; Hsu and Robinson, 2006; Speight, 2007, 2011).

The most versatile modern refinery configurations are the catalytic cracking refinery and the coking refinery (Speight, 2007, 2011). These refineries incorporate all the basic units found in both the topping and hydroskimming refineries, but also feature gas oil conversion plants, such as catalytic cracking and hydrocracking units, olefin conversion plants (i.e. alkylation or polymerization units), and coking units for residuum conversion to reduce or eliminate the production of residual fuels. Modern catalytic cracking and coking refineries produce high outputs of gasoline, with the balance distributed between liquefied petroleum gas, jet fuel, diesel fuel, and a small quantity of coke. Many such refineries also incorporate solvent extraction processes for manufacturing lubricants and petrochemical units with which to recover propylene, benzene, toluene, and xylenes for further processing into polymers.

To convert crude oil into desired products in an economically feasible and environmentally acceptable manner. Refinery process for crude oil are generally divided into three categories:

1. Separation processes, of which distillation is the prime example
2. Conversion processes, of which coking and catalytic cracking are prime examples
3. Finishing processes, of which hydrotreating, to remove sulfur, is a prime example.

Furthermore, as noted elsewhere (Chapter 6), the overall economics or viability of a refinery depends on the interaction of five key elements:

1. The crude slate (i.e. the choice of crude oils accepted by the refinery)
2. The refinery configuration (i.e. the complexity of the refining equipment or refinery configuration)
3. The product slate (i.e. the desired type and quality of products produced)
4. The refinery utilization rates (i.e. portion of the refinery capacity that is utilized)
5. Environmental considerations.

Using more expensive and lighter, sweeter crude oil requires less refinery upgrading, but supplies of light, sweet crude oil are decreasing and the differential between heavier and sourer crudes is increasing. Using cheaper heavy oil means more investment in upgrading processes. Costs and payback periods for refinery processing units must be weighed against anticipated crude oil costs and the projected differential between light and heavy crude oil prices. Crude slates and refinery configurations must take into account the type of products that will ultimately be needed in the marketplace. The quality specifications of the final products are also increasingly important as environmental requirements become more stringent.

The petroleum refining industry provides products that are critical to the functioning of the economy. Virtually all transportation, including land, sea, and air, is fueled by products that are refined from crude oil. Industrial, residential, and commercial activities, as well as electricity generation, use petroleum-based products. Along with volatile changes in crude oil prices, the industry has faced evolving health, safety, and environmental requirements that have changed and multiplied product specifications and required capital investment in refineries.

In fact, the demand for refined products — transportation fuels in particular — has been rising steadily over the past two decades. From 2000 to 2002, consumption of refined petroleum products in the United States peaked at an all-time high of 19.7 million barrels per day.

As the demand for petroleum products has grown, the quality and the performance of petroleum products have changed substantially as a result of environmental regulations and motor vehicle performance requirements. Over the next several years, the industry will be upgrading its facilities for the following purposes:

1. To produce ultra-low-sulfur gasoline and diesel fuel
2. To blend in ethanol and other biofuels (Chapter 9) into gasoline and diesel
3. To reduce the environmental impacts of plant operations.

The 1990s were widely viewed by the industry as a period of unprecedented economic volatility and hardship, characterized by poor profit margins as a result of substantial excess capacity, the increasing cost of compliance with environmental regulations, and unfavorable crude oil price trends. At the same time, consolidation and corporate restructuring have dramatically changed the refining industry in the United States. In addition, most vertically integrated major oil companies have scaled back, shut down, or spun off their process technology development divisions. Many also have shed refineries and retail outlets leading to the emergence of new business models and large independent refining companies.

Since the late 1990s, the industry has undergone significant structural change which might alter its profitability requirements, its ability to provide stable product volumes to the consuming market, as well as its ability to adapt to current and future environmental requirements.

Operations within firms have become more autonomous. In the past, vertically integrated oil companies often managed downstream operations as a means to monetize crude oil production operations; for example, downstream refining operations were often subsidized or financed by the upstream. Currently, refining and marketing operations in the United States are generally managed as stand-alone business units accountable for their own profit-and-loss performance. Disaggregation of business units combined with new management practices has focused attention on obtaining greater returns from existing capital, avoiding unnecessary investment, and cutting costs.

A few significant structural changes characterize the industry. Mergers, acquisitions, and joint ventures have changed the ownership profile of the industry, altering concentration patterns both regionally and nationally. A change in business model from an integrated component, to a stand-alone profit center, has focused attention on requirement of competitive profitability rates from each stage in the production chain if the industry is to remain viable in its current form. Evidence suggests that the new market structure and business model might demand better economic performance from the industry. Regulatory compliance to meet congressionally mandated environmental standards, both on refined products and refinery sites, requires substantial capital investment by refiners, and has resulted in reduced profitability, according to the Energy Information Administration (EIA). To the extent that continued capacity expansion and technological investments are reduced, or not undertaken, because of low rates of return, U.S. dependence on imported refined products might increase, or product markets could be disrupted by shortages and price spikes.

Consolidation and restructuring appear to have had the salutary effect executives intended insofar as there are indications that mid-size and large-size refiners have reduced the per-barrel operating costs by one-third. On the other hand, the elimination of spare downstream capacity generates upward pressure on prices at the pump and produces short-term market vulnerabilities. Disruptions in refinery operations resulting from scheduled maintenance and overhauls or unscheduled breakdowns are more likely to lead to acute supply shortfalls and price spikes, as measured in weeks.

Industry representatives emphasized the regional character of the markets in the United States. This point was brought out in reference to the following:

1. Regional business strategies of refiners
2. The constraints imposed by refinery locations
3. Pipeline infrastructure
4. Fuel specifications.

In addition, market volatility and potential supply shortfalls, should they occur, are most likely to be of regional, not national, scope. The Gulf Coast, Eastern Seaboard, and Southeast are seen as the least-vulnerable regions because of the concentration of refineries and pipelines there and the greater accessibility of imports. The West Coast and Midwest were frequently cited as regions of concern because of regional environmental regulations and the lack of easily accessible alternative supplies, among other issues.

Increased imports of refined products, particularly motor gasoline, combined with growing imports of crude oil could make the U.S. increasingly vulnerable to shocks originating in the world oil market. Importing motor gasoline into the U.S. in appropriate volumes may become increasingly difficult because of the unavailability of world supplies consistent with fuel specification requirements in the Unites States.

Finally, not all investment decisions are driven by refinery economics and market analysis. Refiners also make investment decisions because of voluntary actions or legislative and regulatory requirements. In recent years, governments and industry have directed considerable effort towards reducing the environmental impact of burning fossil fuels. Many of the initiatives have been aimed at providing cleaner fuels. Petroleum refining is a very complicated and capital-intensive industry. New environmental regulations require industry to make additional investments to meet the more stringent standards.

7.4 Profitability

Oil company profits have increased over the past five years and have been very impressive.

Regardless of the relative magnitude of oil company profits, many believe that a large percentage of oil company profits today are unearned in the sense that little or no additional cost or effort was incurred to generate them. Profits from the current price increase are a windfall that does may not rightly belong to producers, if they come at the expense of consumer welfare — or so the argument goes.

Moreover, if excess profits are defined as returns above the normal profits that could be earned through investments in other markets, then the extraction of those profits by governments is possible only through auctions in which participants bid for the right to extract natural resources. Such bids take into account risk and uncertainty about likely outcomes ranging from no discovery to discovery plus low prices to discovery plus high prices.

Proposals to extract profits after the fact are not efficient because they violate investor expectations and change the rules of the game after investments have been made. If investors think that they can keep natural resource profits, they will accept risk because the rewards are potentially quite high. The government reneges when investments are successful, but does not correspondingly help investors when returns are below expectations; investors will reduce their participation in energy markets because profits in energy attract too much political attention relative to investments in other areas of the economy.

Denying investors profits, but allowing them to book losses, amounts to one-way capitalism. Denying the industry the opportunity to make substantial profits when supplies are tight is both unfair and counterproductive, in that it will discourage investment in the oil business, unless their losses are likewise alleviated during low price periods.

The only question that remains is "how much is enough?" in terms of profits realized by oil companies at the expense of the consumer.

7.5 References

Ancheyta, J., and Speight. J.G. 2007. Hydroprocessing of Heavy Oils and Residua. CRC Press, Taylor & Francis Group, Boca Raton, Florida.

Baillie, R., Booth, G., Tse, Y., and Zabotina, T. 2002. Price Discovery and Common Factors Models. Journal of Financial Markets, 5, 305–321.

Booth, G., Lin, J., Martikainen, T., and Tse, Y. 2002. Trading and pricing in upstairs and downstairs stock markets. Review of Financial Studies, 15, 1111–1135.

BP. 2008. BP Statistical Review of World Energy. www.bp.com/statisticalreview.

Harris, L. E. 1991. Stock price clustering and discreteness. Review of Financial Studies, 4: 389–415.

Hasbrouck, J. 1995. One security, many markets: Determining the contribution to price discovery. Journal of Finance, 50: 1175–1199.

Hasbrouck, J. (2002). Stalking the Efficient Price In Market Microstructure Specifications: An Overview. Journal of Financial Markets, 5: 329–339.

Hsu, C.S., and Robinson, P.R. 2006. Practical Advances in Petroleum Processing. Volume 1 and Volume 2. Springer, New York.

Huang, R. 2002. The Quality of ECN and Nasdaq Market Maker Quotes. Journal of Finance, 57: 1285–1319.

Klare, M.T. 2007. Beyond the Age of Petroleum. The Nation, November 12.

Martens, M. 1998. Price discovery in high and low volatility periods: Open outcry versus electronic trading. Journal of International Financial Markets, Institutions and Money, 8, 243–260.

Norman, D.A., and Shin, D. 1991. Price Adjustment in Gasoline and Heating Oil Markets. Research Study No. 060. American Petroleum Institute, Washington, DC. August.

Pirrong, C. 1996. Market Liquidity and Depth on Computerized and Open Outcry Trading Systems: A Comparison Of DTB And LIFFE Bund Contracts. Journal of Futures Markets, 16: 519–544.

Sapp, S. (2002). Price Leadership in the Spot Foreign Exchange Market. Journal of Financial and Quantitative Analysis, 37: 425–428.

Speight, J.G. 2007. The Chemistry and Technology of Petroleum. 4th Edition CRC Press, Taylor and Francis Group, Boca Raton, Florida.

Speight, J.G. 2011. The Refinery of the Future, Gulf Professional Publishing, Elsevier, Oxford, United Kingdom.

Tse, Y., & Zabotina, T. (2001). Transaction Costs and Market Quality: Open Outcry versus Electronic Trading. The Journal of Futures Markets, 21: 713–735.

8

Oil Supply

Crude oil is the product of the burial and transformation of biomass over the last 200 million years and, historically, has faced no equal as an energy source for its intrinsic qualities of extractability, transportability, versatility, and cost. But, the total amount of oil underground is finite, and, therefore, production will one day reach a peak and then begin to decline. Such a peak may be involuntary if supply is unable to keep up with growing demand. Alternatively, a production peak could be brought about by voluntary reductions in oil consumption before physical limits to continued supply growth kick in. Not surprisingly, concerns have arisen in recent years about the relationship between the growing consumption of oil and the availability of oil reserves and the impact of potentially dwindling supplies and rising prices on the world's economy and social welfare. Following a peak in world oil production, the rate of production would eventually decrease and, necessarily, so would the rate of consumption of oil.

Oil can be found and produced from a variety of sources. To date, world oil production has come almost exclusively from what are considered to be conventional sources of oil. While there is no universally agreed upon definition of what is meant by conventional sources, they can be produced using today's mainstream technologies, compared to nonconventional sources that require more complex or more expensive technologies to extract, such as tar sand (oil sand in Canada) (Chapter 4) and oil shale (Chapter 9). Distinguishing between conventional and nonconventional oil sources is important because the additional cost and technological challenges surrounding production of nonconventional sources make these resources more uncertain.

However, this distinction is further complicated because what is considered to be a mainstream technology can change over time. For example, offshore oil deposits were considered to be a nonconventional source 50 years ago; however, today they are considered conventional. This is consistent with the inclusion of tar sand deposits and oil shale in nonconventional sources. Some oil is being produced from these nonconventional sources today. For example, in 2005 Canada produced about 1.6 million barrels per day of oil from oil sands, and Venezuelan production of extra-heavy oil for 2005 was projected to be about 600,000 barrels per day. However, current production from these sources is very small compared with total world oil production.

Crude oil reserves (Chapter 1) are the estimated quantities of oil that are claimed to be recoverable under existing operating and economic conditions. However, because of reservoir characteristics and the limitations of current recovery technologies only a fraction of this oil can be brought to the surface; it is this producible fraction that is considered to be reserves. Crude oil recovery varies greatly from oil field to oil field based on the character of the field and the operating history as well as in response to changes in technology and economics.

Thus, in addition to geological factors, recovery depends on several other factors that are discussed in turn below.

8.1 Physical Factors

In region after region, the story is of ageing fields with depleted oil reserves. For example, Britain's North Sea oil is at its peak, the giant fields in Alaska, the former Soviet Union, Mexico, Venezuela and Norway are all past their peak. Oil supply in the United States now account for less than half the country's needs. There is always a possibility of some big finds in various regions of the world, but once found, development of the fields is still years away. The only producers with an oil resource which may be capable of keeping oil flowing into the world market at a roughly constant level are the Middle East OPEC producers: Saudi Arabia, Iran, Iraq, Kuwait and the United Arab Emirates.

However, most of the oil reserves of Saudi Arabia are held in the large Al Ghawar filed, which has been pumped since 1948 and, not surprisingly, the field is showing signs of exhaustion with its southern end now flooding with water. Saudi Arabia can keep its production roughly constant for another seven to ten years before it too has used up half its total oil resource and rolls over towards depletion. The move will be to smaller fields to produce smaller amounts of crude oil, followed by poor-quality fields, such as the Manifa field).

In the other Gulf States, it is unlikely that Iran could sustain a higher output for the next two decades. Kuwait and one of the emirates, Abu Dhabi, could increase production and may well do so, but their reserves are small, relative to world demand. Only one country, Iraq, has the potential for an increase in output on a scale, which could make a difference. However, Iraq's oil potential has not been fully explored but one estimate is that there are 110 billion barrels of crude oil reserves in Iraq — equal to more than one third of the total resource once possessed by Saudi Arabia. This oil could not be made immediately available, but it is on a scale to keep world oil production rising for a few more years.

Most studies estimate that oil production will peak sometime between now and 2040, although many of these projections cover a wide range of time, including two studies for which the range extends into the next century (Gordon, 2007). Key uncertainties in trying to determine the timing of peak oil are the following:

1. Amount of oil throughout the world
2. Technological, cost, and environmental challenges to produce that oil
3. Political and investment risk factors that may affect oil exploration and production
4. Future world demand for oil. The uncertainties related to exploration and production also make it difficult to estimate the rate of decline after the peak.

Studies that predict the timing of the hypothetical peak in oil production (assuming no further discoveries of giant oil fields) use different estimates of how much oil remains in the ground, and these differences explain some of the wide ranges of these predictions. Estimates of how much oil remains in the ground are highly uncertain because much of this data is self-reported and unverified by independent auditors; many parts of the world have yet to be fully explored for oil; and there is no comprehensive assessment of oil reserves from nonconventional sources. This uncertainty surrounding estimates of oil resources in the ground comprises the uncertainty surrounding estimates of proven reserves as well as uncertainty surrounding expected increases in these reserves and estimated future oil discoveries.

According to current estimates, more than three-quarters of the world's oil reserves are located in OPEC countries. The bulk of OPEC oil reserves is located in the Middle East, with Saudi Arabia, Iran and Iraq contributing 41.8% to the OPEC total. OPEC member countries have made significant contributions to their reserves in recent years by adopting best practices in the industry. As a result, OPEC proven reserves currently stand at 934.7 billion 900 billion barrels while non-OPEC reserves, including Russia, are on the order of 303.1 billion barrels (BP, 2008).

There has been surprise at the OPEC estimates of proven reserves (Campbell and Laherrère, 1998) since OPEC estimates increased sharply in the 1980s, corresponding to a change in OPEC's quota rules that linked a member country's production quota in part to its remaining proven reserves. Indeed, companies that are not subject to the federal securities laws in the United States and their related liability standards, include companies wholly owned by various OPEC member countries where the majority of reserves are located.

In addition, many OPEC countries' reported reserves remained relatively unchanged during the 1990s, even as they continued high levels of oil production. For example, estimates of reserves in Kuwait were unchanged from 1991 to 2002, even though the country produced more than 8 billion barrels of oil over that period and did not make any important new oil discoveries. The potential disbelief in the data reported by OPEC is problematic with respect to predicting the timing of a peak in oil production because OPEC holds most of the world's current estimated proven oil reserves. On the basis of reserve estimates as of December 2007, of the approximately 1.24 billion barrels of proven oil reserves worldwide, 934.7 billion barrels (75.5%) of oil are located in the OPEC countries, compared with 29.4 billion barrels (2.4%) in the United States (BP, 2008).

The United States Geological Survey provides oil resources estimates, which are different from proved reserves estimates. Oil resources estimates are significantly higher because they estimate the world's total oil resource base, rather than just what is now proven to be economically producible. Estimates of the resource by the United States Geological Survey base include past production and current reserves, as well as the potential for future increases in current conventional oil reserves (often referred to as reserves growth) and the amount of estimated conventional oil that has the potential to be added to these reserves. Estimates of reserves growth and those resources that have the potential to be added to oil reserves are important in determining when oil production may peak.

However, the estimating of these potential future reserves is complicated by the fact that many regions of the world have not been fully explored and, as a result, there is limited information. For example, in 2000, the United States Geological Survey provided a mean estimate of 732 billion barrels that have the potential to be added as newly discovered conventional oil, with as much as 25 percent from the Arctic, including Greenland, Northern Canada, and the Russian portion of the Barents Sea. However, relatively little exploration has been done in this region, and there are large portions of the world where the potential for oil production exists, but where exploration has not been done. According to the United States Geological Survey there is less uncertainty in regions where wells have been drilled, but even in the United States, one of the areas that has seen the greatest exploration, some areas have not been fully explored, as illustrated by the recent discovery of a potentially large oil field in the Gulf of Mexico.

Limited information on oil-producing regions worldwide also leads the United States Geological Survey to base its estimate of reserves growth on how reserves estimates have grown in the United States. However, some experts criticize this methodology; they believe such an estimate may be too high because the United States experience overestimates increases in future worldwide reserves. In contrast, EIA believes the estimate by the United States Geological Survey estimate may be too low. In 2005, the United States Geological Survey released a study showing that its prediction of reserves growth has been in line with the world's experience from 1996 to 2003. Given such controversy, uncertainty remains about this key element of estimating the amount of oil in the ground. In 2000, the most recent full assessment of the world's key oil regions by the United States Geological Survey provided a range of estimates of remaining world conventional oil resources. The mean of this range was at about 2.3 trillion barrels, comprising about 890 billion barrels in current reserves and 1.4 trillion barrels that have the potential to be added to oil reserves in the future.

Further contributing to the uncertainty of the timing of a peak is the lack of a comprehensive assessment of oil from nonconventional sources. For example, the three key sources of oil estimates, *Oil and Gas Journal, World Oil,* and the United States Geological Survey, do not generally include oil from nonconventional sources. Yet oil from nonconventional sources is exists in large quantities. For example, that oil from nonconventional sources, such as the Alberta tar sands, and similar material in the Venezuelan deposits, as well as oil shale in the United States, accounts for as much as 7 trillion barrels of oil, which could greatly delay the onset of a peak in production. However, challenges facing this production (Chapter 9) indicate that the amount of nonconventional oil that will eventually be produced is highly uncertain.

Despite this apparent uncertainty, development and production of oil (synthetic crude oil) from the Alberta tar sands and Venezuelan extra-heavy oil production are under way now and the information available indicates that these sources will be effective in the near future (Chapter 5).

8.2 Technological Factors

In region after region, the story is of ageing fields, of the wrong sort of oil, of nitrogen being pumped into wells to keep up the flow, of new areas turning out to be dry, and oil is at its peak now. The giant fields in Alaska, the former Soviet Union, Mexico, Venezuela and Norway are all past their peak. There is a possibility of some big finds off the coast of west Africa, but their development is still years into the future, and they are not on a scale capable of making a difference. The only producers with an oil resource that may be capable of keeping oil flowing into the world market at a roughly constant level are the Middle East OPEC five: Saudi Arabia, Iran, Iraq, Kuwait and the United Arab Emirates (Fleming, 2000). However, it is also difficult to project the timing of a peak in oil production because technological, cost, and environmental challenges make it unclear how much oil can ultimately be recovered from proven reserves, hard-to-reach locations, and nonconventional sources.

The world's total production of liquid fuels from unconventional resources in 2006 was 2.8 million barrels per day, equal to about 3 percent of total liquids production. Production from unconventional sources included 1.2 million barrels per day from tar sand in Canada, 600,000 barrels per day from very heavy oils in Venezuela, and 320,000 barrels of ethanol per day in the United States (Annual Energy Outlook, 2008).

To increase the recovery rate from oil reserves, companies turn to enhanced oil recovery technologies, which United States Department of Energy reports has the potential to increase recovery rates from 30 to 50 percent in many locations.These technologies include injecting steam or heated water; gases, such as carbon dioxide; or chemicals into the reservoir to stimulate oil flow and allow for increased recovery. Opportunities for enhanced oil recovery have been most aggressively pursued in the United States, where enhanced oil recovery technologies currently contribute approximately 12 percent to U.S. production, and carbon dioxide EOR alone is projected to have the potential to provide at least 2 million barrels per day by 2020. However, technological advances, such as better seismic and fluid-monitoring techniques for reservoirs during an enhanced oil recovery injection, may be required to make these techniques more cost-effective. Furthermore, enhanced oil recovery technologies are much costlier than the conventional production methods used for the vast majority of oil produced. Costs are higher because of the capital cost of equipment and operating costs, including the production, transportation, and injection of agents into existing fields and the additional energy costs of performing these tasks. Finally, enhanced oil recovery technologies have the potential to create environmental concerns associated with the additional energy required to conduct an enhanced oil recovery injection and the greenhouse gas emissions associated with producing that energy, although EIA has stated that these environmental costs may be less than those imposed by producing oil in previously undeveloped areas. Even if sustained high oil prices make enhanced oil recovery technologies cost-effective for an oil company, these challenges and costs may deter their widespread use.

The timing of peak oil is also difficult to estimate because new sources of oil could be increasingly more remote and costly to exploit, including offshore production of oil in deep-water and ultra deep-water. Worldwide, industry analysts report that deep-water (depths of 1,000 to 5,000 feet) and ultra deep-water (5,000 to 10,000 feet) drilling efforts are concentrated offshore in Africa, Latin America, and North America, and capital expenditures for these efforts are expected to grow through at least 2011. In the United States, deep-water and ultra deep-water drilling, primarily in the Gulf of Mexico, could reach 2.2 million barrels per day in 2016, according to EIA estimates. However, accessing and producing oil from these locations present several challenges. At deep-water depths, penetrating the earth and efficiently operating drilling equipment is difficult because of the extreme pressure and temperature. In addition, these conditions can compromise the endurance and reliability of operating equipment. Operating costs for deep-water rigs are 3.0 to 4.5 times more than operating costs for typical shallow water rigs. Capital costs, including platforms and underwater pipeline infrastructures, are also greater. Finally, deepwater and ultra deep-water drilling efforts generally face similar environmental concerns as shallow water drilling efforts, although some deep-water operations may pose greater environmental concerns to sensitive deep-water ecosystems.

It is unclear how much oil can be recovered from nonconventional sources. Recovery from these sources could delay a peak in oil production or slow the rate of decline in production after a peak. There is disagreement and much speculation about the decline rate and disagreement concerning the significance of the role these nonconventional sources will play in the future. However, IEA estimates of oil production have conventional oil continuing to comprise almost all of production through 2030.

The development and widespread adoption of technologies to displace oil will take time and effort, an imminent peak and sharp decline in oil production could have severe consequences. The technologies we examined currently supply the equivalent of only about 1% of U.S. annual consumption of petroleum products, and the United States Department of Energy projects that even under optimistic scenarios, these technologies could displace only the equivalent of about 4% of annual projected U.S. consumption by around 2015. If the decline in oil production exceeded the ability of alternative technologies to displace oil, energy consumption would be constricted, and as consumers competed for increasingly scarce oil resources, oil prices would sharply increase. In this respect, the consequences could initially resemble those of past oil supply shocks, which have been associated with significant economic damage. For example, disruptions in oil supply associated with the Arab oil embargo of 1973–74 and the Iranian Revolution of 1978–79 caused unprecedented increases in oil prices and were associated with worldwide recessions. In addition, a number of studies we reviewed indicate that most of the U.S. recessions in the post-World War II era were preceded by oil supply shocks and the associated sudden rise in oil prices.

Ultimately, the consequences of a peak in oil production and permanent decline in oil production could be even more prolonged and severe than those of past oil supply shocks. Even then the decline rate is the subject of speculation. The only certainty is that the decline rate is happening. The most important variable is the amount of oil left in the reservoirs, but, even then, this is subject to debate and error leaving the decline rate for fields in production difficult to assess (Eagles, 2006; Gerdes, 2007; Jackson, 2007). At best, generalities can be calculated. For example, for current fields in production a low decline rate of 2% per year would result in peak oil around 2018 while a more moderate decline rate at 4.5% per year would result in peak oil around 2015 and a high decline rate of 8% per year would result in peak oil around 2010.

In addition, because the decline would be neither temporary nor reversible, the effects would continue until alternative transportation technologies to displace oil became available in sufficient quantities at comparable costs. Furthermore, because oil production could decline even more each year following a peak, the amount that would have to be replaced by alternatives could also increase year by year.

8.3 Economic Factors

The economics of oil is dominated by the proximity of oil production to the so-called peak oil. The decline in the discovery of oil over the past decades means that production, too, must decline. While there may be short-term fluctuations in oil prices, when there is a belief that the price in the future will be higher than the price for delivery straight away, there will be a flurry of market activity to buy up short-term contracts. This will bid up prices, leading to a new equilibrium at a much higher level and persuading producers that the longer they leave the oil in the ground, the better the price they will get (Fleming, 2000).

In addition, foreign investment in the oil sector could be necessary to bring oil to the world market but many countries have restricted foreign investment. Lack of investment could hasten a peak in oil production because the proper infrastructure might not be available to find and produce oil when needed, and because technical expertise may be lacking. The important role foreign investment plays in oil production is illustrated in Kazakhstan where opening the energy sector to foreign investment in the early 1990s led to a doubling in oil production between 1998 and 2002. In addition, we found that direct foreign investment in Venezuela was strongly correlated with oil production in that country, and that when foreign investment declined between 2001 and 2004, oil production also declined.

The lack of technical expertise can lead to less sophisticated drilling techniques that actually reduce the ability to recover oil in more complex reservoirs. For example, according to industry officials, some Russian wells have difficulties with high water cut (a high ratio of water to oil), making oil difficult to get out of the ground at certain price levels for crude oil. This water cut problem stems from not using technically advanced methods when the wells were initially drilled. The Venezuelan national oil company, PDVSA, lost technical expertise when it fired thousands of employees following a strike in 2002 and 2003. In contrast, other national oil companies, such as Saudi Aramco, are widely perceived to possess considerable technical expertise.

A high proportion of approximately 85% of the world's proven oil reserves are in countries with medium-to-high investment risk or where foreign investment is prohibited. For example, over one-third of the world's proven oil reserves lie in only five countries: China, Iran, Iraq, Nigeria, and Venezuela. All of these countries have a high likelihood of seeing a worsening investment climate. hree countries with large oil reserves (Saudi Arabia, Kuwait, and Mexico) prohibit foreign investment in the oil sector, and most major oil-producing countries have some type of restrictions on foreign investment. Furthermore, according to the United States

Department of Energy some countries that previously allowed foreign investment, such as Russia and Venezuela, appear to be reasserting state control over the oil sector.

National oil companies may have additional motivations for producing oil, other than meeting consumer demand. For instance, some countries use some profits from national companies to support domestic socioeconomic development, rather than focusing on continued development of oil exploration and production for worldwide consumption. Given the amount of oil controlled by national oil companies, these types of actions have the potential to result in oil production that is not optimized to respond to increases in the demand for oil.

While current high oil prices may encourage development and adoption of alternatives to oil, if high oil prices are not sustained, efforts to develop and adopt alternatives may fall by the wayside. The high oil prices and fears of running out of oil in the 1970s and early 1980s encouraged investments in alternative energy sources, including synthetic fuels made from coal, but when oil prices fell, investments in these alternatives became uneconomic. More recently, private sector interest in alternative fuels has increased, corresponding to the increase in oil prices, but uncertainty about future oil prices can be a barrier to investment in risky alternative fuels projects. Also, interest in fuel efficiency tends to increase as gasoline prices rise and decrease when gasoline prices fall.

Moreover, the economic principles, which explain how a market economy works, tend to break down when applied to natural resources such as oil. In fact, there are two ways in which the principles of market economics do not apply to crude oil: (1) the current price of oil has virtually no influence on the rate at which it is discovered and (2) the rules of supply and demand do not always hold and a rise in the price of crude oil does not always lead to an increase in production.

There is still a large quantity of oil in the ground but what really matters is not how much remains, but the turning point at which the flow of oil hits its peak and starts to turn down. Furthermore, the world as a whole currently uses at least 30% more oil now than it did in 1970, and the fact that its consumption of gas has risen many times over does not mean that there is less dependency on oil. It does mean that the world has become more dependent on gas. The move to alternate sources of energy is touted as the savior of the energy-consuming countries. But, there are time lags.

There are assumptions that renewable sources of energy will come on stream just in time to take over from oil. All that is required is to wait for the price signal insofar as when oil becomes more expensive, energy from alternative sources will be immediately available and solve the energy problems. The unanswered question related to fiscal actions taken by the OPEC nations that might deter governments from tackling and funding the development of the more expense alternate energies. One also has to wonder if the politicians at various levels of government are willing to tell their respective constituents that, for example, gasoline from a renewable source will cost more than gasoline from petroleum at the risk of losing votes and a their respective seats in government.

In addition, the development of alternative energy sources to fill the void left by the end of petroleum will take time. The development of renewable energy systems needs to be supported by decisive, well-coordinated action by governments, in sustained multi-decade programs. Only then will renewable sources be poised to supplement petroleum. Obviously, if the oil-consuming nations wait for the market to give the price signal that renewable forms of energy should now be developed, the effort we start 25 years too late and there will be a destabilizing energy gap.

The economics of oil is now dominated by its close proximity to the output peak. The steep decline in the discovery of oil since 1965 means that production must eventually decline beyond a point of no return. Recent rises in oil prices suggest that the very high prices associated with the summer of 2008 and the fluctuations to more moderate prices are temporary. The tension between demand and the reduced growth in supply can be expected to raise prices again. When the consumers begin to believe that the price in the future will be higher than the price for delivery, there will be a rush to buy up short-term contracts leading to new price equilibrium, but at a much higher level. This may even be a signal to producers that the longer they leave the oil in the ground, leading to a higher price for oil resulting in a stalemate between high prices and flattened demand before supply collapses into a decline.

The consequences of such an economic turndown will affect the two main purposes for which oil is used: food and transport.

Agriculture in the oil-consuming world is heavily dependent on crude oil, but this dependency could be reduced by a switch to renewable sources of energy, such as solar and wind power. However, that invokes the concept of a multi-year — perhaps two decades — transition period. In short, the dependency played by crude oil in the provision of food is almost non-negotiable and will force consumers to continue to buy oil to the limit of their resources.

The need for crude oil to keep transport moving in the United States will remain real and unavoidable for at least 25 years. The US economy is organized on the assumption of cheap, long-distance transport. As is already evident, in these times crude oil supply is in decline, the United States is increasing its consumption. With the daily lives of consumers locked into dependency on road transport, it is already evident that the consumers are straining to cope with prices and there are already signs of economic destabilization.

Governments are now in a dilemma and have to acknowledge this problem with the implications that the economics of oil must be taken seriously. If the barrier between rational thinking and institutional complacency continues to hold, the energy future will be bleak.

As governments eventually acknowledge the problem of oil supply and demand and the various economic consequences, the first step should be to establish a dialogue with their public and no to pontificate. The government needs to inform the public of the proposed actions since, by then (perhaps even now), the public will require political leaders who present themselves as moving towards economic stability.

For example, governments must organize the quest for alternative energy sources and conservation technologies. However, an energy system based on renewable sources is not capable at this time of providing the energy needed for transport on the present scale but, nevertheless, a meaningful (not a token) start must be made and available technologies, such as solar and wind technologies, need to be applied urgently.

In the interests of economic stability, there may be a case, despite the climate change implications, for the United States and other coal-rich countries to move to the use of coal as a source of transportation fuels (Chapter 9). The use of coal can help such countries buy time in the form of a coal-based energy system that will allow development future sources of alternate energy. During this time, the increased emissions of carbon dioxide will be partly compensated for by a decline in emissions from oil.

8.4 Geopolitical Factors

The true picture of oil supply may never be known. Difficult as it is because of a variety of factors, reporting data on oil production or oil reserves is a political act (Laherrère, 2001). The SEC, to satisfy bankers and shareholders, obliges the oil companies listed on the US stock market to report only proved reserves and to omit probable reserves that are reported in the rest of the world. This practice of reporting only proved reserves can lead to a strong reserve growth since 90% of the annual reserves oil addition come from revisions of old fields, showing that the assessment of the fields was poorly reported. This reserve growth of conventional oil reserves is incorrectly attributed to technological progress. Technical data, on which development decisions are based, exist but they are confidential. Reporting of production is not much better and may give a false impression of oil abundance (Simmons 2000).

The pricing for crude oil is generally politically determined. Treated as a commodity, and hence traded for profit by private corporations and producing nations, crude oil has long been thought of as subject to the free interplay of competitive forces, but is actually influenced by political considerations. The modern oil corporation, with assets greater than those of most countries, has maintained basic surveillance over all energy development. A brief summary of the history of oil (Chapter 1) will show the manner in which politics enters into the oil picture.

The modern era of oil production and the ensuing age of petropolitics began on August 27, 1859, when Edwin L. Drake drilled the first successful oil well 69 feet deep near Titusville in northwestern Pennsylvania. Just five years earlier, the invention of the kerosene lamp had ignited intense demand for oil. By drilling an oil well, Drake had hoped to meet the growing demand for oil for lighting and industrial lubrication. Drake's success inspired hundreds of small companies to explore for oil. In 1860, world oil production reached 500,000 barrels; by the 1870s production soared to 20 million barrels annually. In 1879, the first oil well was drilled in California; and in 1887, in Texas. As production boomed, prices fell and oil industry profits declined, until in 1882, John D. Rockefeller had devised a solution to the problem of competition in the oil fields: the Standard Oil Trust. This brought together 40 of the nation's leading refiners and through its control of refining, Standard Oil was able to control the price of oil.

During the early 20th century, oil production continued to climb. By 1920, oil production reached 450 million barrels, prompting fear that the nation was about to run out of oil. Government officials predicted that the nation's oil reserves would last just ten years. Up until the 1910s, the United States produced between 60 and 70% of the world's oil supply. As fear grew that American oil reserves were dangerously depleted, the search for oil turned worldwide. Oil was discovered in Mexico at the beginning of the 20th century, in Iran in 1908, in Venezuela during World War I, and in Iraq in 1927. Many of the new oil discoveries occurred in areas dominated by Britain and the Netherlands: in the Dutch East Indies, Iran, and British mandates in the Middle East. By 1919, Britain controlled 50% of the world's proven oil reserves.

After World War I, a bitter struggle for control of world oil reserves erupted. The British, Dutch, and French excluded American companies from purchasing oil fields in territories under their control. Congress retaliated in 1920 by adopting the Mineral Leasing Act, which denied access to American oil reserves to any foreign country that restricted American access to its reserves. The dispute was ultimately resolved during the 1920s when American oil companies were finally allowed to drill in the British Middle East and the Dutch East Indies.

The fear that American oil reserves were nearly exhausted ended abruptly in 1924, with the discovery of enormous new oil fields in Texas, Oklahoma, and California. These discoveries, along with production from new fields in Mexico, the Soviet Union, and Venezuela, combined to drastically depress oil prices. By 1931, with crude oil selling for 10 cents a barrel, domestic oil producers demanded restrictions on production in order to raise prices. Texas and Oklahoma passed state laws and stationed militia units at oil fields to prevent drillers from exceeding production quotas. Despite these measures, prices continued to fall.

In a final bid to solve the problem of overproduction, the federal government stepped in. Under the National Recovery Administration, the federal government imposed production restraints, import restrictions, and price regulations. After the

Supreme Court declared the NRA unconstitutional, the federal government imposed a tariff on foreign oil.

During World War II, the oil surpluses of the 1930s quickly disappeared. Six billion of the seven billion barrels of petroleum used by the allies during the war came from the United States. Public officials again began to show concern that the United States was running out of oil.

On the other hand, world oil prices were so low that Iran, Venezuela, and Arab oil producers banded together in 1960 to form OPEC (the Organization of Petroleum Exporting Countries) cartel, to negotiate for higher oil prices. By the early 1970s, the United States depended on the Middle East for a third of its oil and foreign oil producers were finally in a position to raise world oil prices. The oil embargo of 1973 and 1974, during which oil prices quadrupled, and the oil crisis of 1978 and 1979, when oil prices doubled, underscored the vulnerability of the United States to foreign producers.

The oil crises of the 1970s had an unanticipated side effect. Rising oil prices stimulated conservation and exploration for new oil sources. As a result of increasing supplies and declining demand, oil prices fell from $35 a barrel in 1981 to $9 a barrel in 1986. The sharp slide in world oil prices was one of the factors that led Iraq to invade neighboring Kuwait in 1990 in a bid to gain control over 40 percent of Middle Eastern oil reserves.

On the other hand, oil producers operating outside of the OPEC cartel are responsible for producing 60% of the world's oil and face increasing production hurdles. However, many of the non-OPEC producers have older, less productive wells, rising costs for new projects, and in some cases rising domestic demand that may cut into exports. Higher prices have made difficult oil projects more lucrative, leading to increases in unconventional oil production, but that could change. Declines in non-OPEC production come at a time when investment in new oil production is more difficult because of tightening credit markets, oil price volatility, and resource nationalism. While a few producers are expected to offset some of these declines, new production is coming online more slowly than originally projected and the world is entering a period of growing demand amidst tightening supplies (NPC, 2007).

Five of the world's 15 largest oil producers are outside of OPEC; as of 2008, those countries are Russia, the United States, China,

Mexico, Canada, Norway, and Brazil http://tonto.eia.doe.gov/country/index.cfm. Non-OPEC nations produced approximately 60% of total production for the year.

However, many non-OPEC producers are faced with wells that are quickly depleting. Some major producers, such as the United States, Mexico, and Norway, have experienced a decline in production in recent years but, on the other hand, non-OPEC production, although declining, has been bolstered by the significant increases in production from Brazil, Canada, Russia, and other former Soviet states (BP, 2008). In 2009, non-OPEC production is expected to increase by 1.5 million barrels per day.

8.5 Peak Oil

After all of the above issues have been taken into account, there remains an issue that is in actual fact a combination of those above. And that is peak oil, the point in time when the maximum rate of global petroleum recovery realized, after which the rate of production enters terminal decline (Deffeyes, 2002, 2005).

However, peak oil is not about running out of oil but the peaking and subsequent decline of the production rate of oil. Many observers have interpreted peak oil to mean that the world is running out of oil at the time of the peak and the downside (the decreased production side of the curve) is often thought to be a precipice down which the oil-consuming nations (the oil-importing nations) speed to oblivion.

8.5.1 Peak Oil Theory

M. King Hubbert created *the Hubbert peak theory* in 1956 to accurately predict that oil production in the United States would reach a maximum between 1965 and 1970 (Hubbert, 1956). The concept is based on the observed production rates of individual oil wells, and the combined production rate of a field of related oil wells. The aggregate production rate from an oil field over time appears to grow exponentially until the rate peaks and then declines, sometimes rapidly, until the field is depleted.

According to the Hubbert postulate, the production rate of a limited resource, such as crude oil, will follows a bell-shaped curve representing increase in production following by a diminished rate of production, then a decline in production. Furthermore, the Hubert model has been shown to be descriptive of the peak and decline of production from oil fields in many countries (Brandt, 2007).

Hubbert predicted in 1974 that peak oil would occur (if the trends prior to 1974 continued) in 1995 (http://www.hubbertpeak.com/hubbert/natgeog.htm). However, in the late 1970s and early 1980s, global oil consumption actually dropped due to the shift to energy–efficient cars and the shift to electricity and natural gas for heating, then rebounded to a lower level of growth in the mid 1980s. Thus, oil production did not peak in 1995, and has climbed to more than double the rate initially projected. This underscores the fact that the only reliable way to identify the timing of peak oil may be in retrospect. However, predictions have been refined through the years as up-to-date information becomes more readily available, such as new reserve growth data. Predictions of the timing of peak oil include the possibilities that it has recently occurred, that it will occur shortly, or that a plateau of oil production will sustain supply for several decades. None of these predictions dispute the peaking of oil production, but disagree only on when it will occur.

However, the argument for peak oil is not presented in the context of a systematic evaluation of available data. The concept of the Hubbert-derived bell was a start in terms of a necessary examination of a finite resource and has been accepted as fact by many who fail to recognize that the potential for a skewed bell curve through the potential of technological advances and the impact of heavy oil and tar sand bitumen; the original concept was developed in the mid-to-late 1950s. Indeed, the post-peak reservoir decline curve assumptions are rebutted by observation that the geometry of typical oilfield production profiles is often distinctly asymmetrical and does not generally show a precipitous mirror-image decline in production after an apparent peak.

In fact, the peak oil theory, because it has been approached as an emotional issue, is causing confusion leading to inappropriate actions that turn attention away from the real issues. Oil is critical to the global economy and emotional outbursts should not replace careful analysis about the very real challenges with delivering liquid fuels to meet the needs of growing economies. However, the global decline will begin by 2020 or later, and assume major investment in energy alternatives will occur before a crisis, without requiring major changes in the lifestyle of heavily oil-consuming nations (Jackson, 2006).

Historically, this is the fifth time that the world is said to be running out of oil. Each time (whether it was the shortage of gasoline at the end of World War I or the gasoline shortage of the 1970s) advances in technology have managed to offset the decline of available oil. In fact, there is sufficient crude oil remaining in the ground that will become available as more technological advances are explored.

Conventional crude oil reserves include all crude oil that is technically possible to produce from reservoirs through a well bore, using primary, secondary, improved, enhanced, or tertiary methods; this does not include liquid fuels produced from tar sand, coal, oil shale, and gas-to-liquid process (Chapter 9) (Speight, 2008). Oil reserves (Chapter 2) are classified as proven, probable, and possible. Proven reserves are generally intended to have at least 90% or 95% certainty of containing the amount specified. Probable reserves have an intended probability of 50%, and the possible reserves have an intended probability of 5% or 10%. Current technology is capable of extracting about 40% of the oil from most wells and it is likely that future technology will make further extraction possible.

The global resource base of conventional and unconventional oils, including historical production of 1.08 trillion barrels and yet-to-be-produced resources, is 4.82 trillion barrels and likely to grow (Jackson, 2006) — the operative word being likely — but subject to advances in technology making unavailable residual oil available, likely has the distinct chance of becoming real; and such oil does not include the so-called yet-to-be-discovered oil (Chapter 9).

However, there is some question about the validity of the reserves estimates given by oil-producing (OPEC) countries (Campbell and Laherrère, 1998; Cohen, 2007). Many of the so-called reserves may actually be inflated and are, in fact, resources; they are not delineated and are inaccessible, so are unavailable for production.

Nevertheless, it is thought that global crude oil production will follow an undulating plateau for one or more decades before declining slowly and the slope of decline will be more gradual and much less steep as the rapid rate of increase of oil discoveries (Figure 8.1) (Jackson, 2006; Cohen, 2007). It is anticipated that increasing oil prices to support the current production plateau for another two decades, but beyond that, at a time when the world's oil fields will be severely depleted (Campbell 2004a, 2004b, 2005), it is difficult to predict the actual events.

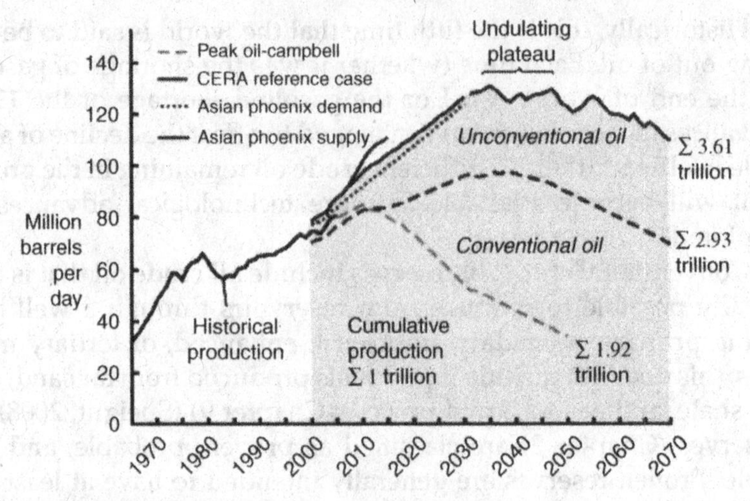

Figure 8.1 Illustration of the undulating plateau (skewed bell curve) of oil production.
*Source: Cambridge Energy Research Associates.

Furthermore, this production pattern (for the original Hubbert prediction) does not include the impact of heavy oil field and tar sand bitumen (Chapters 2 and 5) on the availability of oil by conversion of the latter to synthetic crude oil. In addition, technologies for the production of gas-related liquids (condensate and natural gas liquids), gas-to-liquids (GTL), and coal-to-liquids (CTL) (Chapter 9) will also be needed to bridge the potential gap in the supply of liquid fuels.

In short, before the panic that the world is running out of oil, continues, all of the options must be explored on a logical and systematic basis.

On the other hand, the rigid belief in undiscovered oil resources (Chapter 9) should not detract from the depletion of known oil reserves. The importance of tar sand resources in Canada and similar resources of the Orinoco tar belt and oil shale resources (Chapter 9) provide a global inventory which, when developed, will allow productive capacity of hydrocarbon-based fuels to continue to expand and continue well into the foreseeable future.

Pessimistic predictions of future oil production operate on the thesis that either the peak has already occurred or that it will occur shortly and predictions vary greatly as to what exactly these negative effects would be.

The demand side of peak oil is concerned with the consumption over time, and the growth of this demand. For example, world crude oil demand has grown at almost 2% per year from 1994 and is projected to increase more than 30% over current levels by 2030 (i.e., 118 million barrels per day from 86 million barrels), due in large part to increases in demand from the transportation sector (EIA, 2007).

Heavy crude oil and unconventional sources of liquid fuels (such as tar sands and oil shale) are not counted as part of oil reserves. However, oil companies can book them as proven reserves after opening a mine for tar sand and, thence, bitumen recovery, or thermal facility for recovery of modified bitumen. These unconventional sources are not as efficient to produce, however, requiring extra energy to refine, resulting in higher production costs more greenhouse gas emissions per barrel equivalent. While the energy used, resources needed, and environmental effects of extracting unconventional sources has traditionally been prohibitively high, the three major unconventional oil sources being considered for large scale production are the heavy oil and bitumen in the Orinoco Belt of Venezuela), the tar sand at Athabasca in Alberta), and the oil shale of the Green River formation of Colorado, Utah, and Wyoming in the United States. Taken together, these resource occurrences are approximately equal to the Identified Reserves of conventional crude oil accredited to the Middle East (Speight, 2008).

However, despite the large quantities of oil available in nonconventional sources, limitations on production prevent them from becoming an immediate replacement for conventional crude oil and peak oil will still be a reality. Even under highly optimistic assumptions, production of synthetic crude oil from the Alberta tar sand could reach 5 million barrels per day by 2030 provided there was an immediate program for development (Söderbergh *et al.*, 2007) making it possible that within the foreseeable future the world's extra oil supply will likely come from unconventional sources.

8.5.2 Effects and Consequences of Peak Oil

The widespread use of fossil fuels has been one of the most important stimuli of economic growth and has allowed the consumption of energy at a greater rate than it is being replaced (Yergin, 1993). Thus, the impact of peak oil will depend heavily on the rate of decline and the timely development and adoption of effective alternative sources of energy (Chapter 9). If alternatives are not forthcoming in a timely manner, the multitude of products from crude oil would become scarce and expensive, which, at the very least, could lower living standards in developed and developing countries alike, and in the worst case lead to worldwide economic collapse. With increased tension between countries over dwindling oil supplies, political situations may change dramatically and inequalities between countries and regions may become exacerbated.

The peaking of world oil production presents the U.S. and the world with an unprecedented risk management problem (Hirsch *et al.*, 2005). As peaking is approached, liquid fuel prices and price volatility will increase dramatically, and, without timely mitigation, the economic, social, and political costs will be unprecedented. Viable mitigation options exist on both the supply and demand sides, but to have substantial impact, they must be initiated more than a decade in advance of peaking.

World oil peaking is going to happen, but whether it will occur slowly or abruptly is not certain. There will be an adverse effect on global economies, particularly in those nations most dependent on oil. However, the adoption of alternate technologies to supplant the deficit in oil production will require a substantial time period on the order of ten-to-twenty years. If actions are taken before the onset of peak oil, economic upheaval is not inevitable and the problems can be solved with existing technologies but government intervention is necessary.

The issue of peak oil can be overcome by the following factors:

1. Delay until world oil production peaks before taking crash program action — this would leave the world with a significant liquid fuel deficit for more than two decades
2. Initiate an urgent program for alternate fuel source development 10 years before world oil peaking — this would leave a liquid fuels shortfall for several years (perhaps a decade) after peak oil
3. Initiate an urgent program for alternate fuel source development 20 years before peak oil — this would offer the possibility of avoiding a world liquid fuels shortfall for the forecast period.

The development of such programs will reduce the consumption of traditional petroleum sources as well as affect the timing of peak oil and the shape of the Hubbert curve.

8.6 The Impact of Heavy Oil and Tar Sand Bitumen

Tar sand, called oil sand in Canada, is a deposit of sand containing bitumen that is so viscous it will not flow and is immobile in the deposit unless heated (Chapter 2). As described elsewhere (Chapter 1), the definition of tar sand bitumen is derived from the definition of tar sand that has been defined by the United States government (FE-76-4):

> [Tar sands]...the several rock types that contain an extremely viscous hydrocarbon which is not recoverable in its natural state by conventional oil well production methods including currently used enhanced recovery techniques. The hydrocarbon-bearing rocks are variously known as bitumen-rocks oil, impregnated rocks, oil sands, and rock asphalt.

By inference, heavy oil is a resource that can be recovered in its natural state by conventional oil well production methods, including currently used enhanced recovery techniques. Under this definition, the Cold Lake oil being covered by cyclic steam stimulation (one of the methods enumerated in the US Congressional Record would not be classified as tar sand bitumen but as heavy oil. While most conventional crude oil flows naturally or is pumped from the ground, tar sand bitumen must be recovered by mining or recovered in-situ before being converted into an upgraded crude oil that can be used by refineries to produce gasoline and diesel fuels. Alberta, Canada, contains at least 85% of the world's proven oil sands reserves. In the United States, tar sand deposits are located in Alabama, Alaska, California, Texas, and Utah.

Heavy oils are dense, viscous oils that generally require advanced production technologies, such as enhanced oil recovery, and substantial processing to be converted into petroleum products. Heavy oils differ in their viscosities and other physical properties, but advanced recovery techniques, such as enhanced oil recovery, are required for both types of oil.

Heavy oil can be found in Alaska, California, and Wyoming, and exist in other countries besides the United States and Venezuela. However, like production from oil sands, heavy oil production in the United States presents environmental challenges in its consumption of other energy sources, which contributes to greenhouse gases, and potential groundwater contamination from the injectants needed to thin the oil enough so that oil will flow through pipes. Canadian tar sand and heavy oil, and Venezuelan tar sand and heavy oil are expected to represent cumulatively at least 3 trillion barrels of higher cost crude oil resources. In fact, the known heavy oil reserves in Venezuela represent almost 90% of the proven heavy oil reserves in the world. Venezuelan production of heavy oil approaching one million barrels per day per day is projected to be sustained at this rate through 2040.

Worldwide production of tar sand bitumen, largely from Alberta, contributed approximately 1.6 million barrels of oil per day, and over the next 10 to 12 years output is expected to increase to 3.5 million barrels per day by 2030. More refiners will begin investing to process it and come to depend on the Synbit and Dilbit for a significant part of their supply (Pavone, 2009). However, production of liquid fuels from tar deposits presents significant environmental challenges. The production process uses large amounts of natural gas, which generates greenhouse gases when burned. In addition, large-scale production of bitumen and synthetic fuel from tar sand requires significant quantities of water, typically produce large quantities of contaminated wastewater, and alter the natural landscape. These challenges may ultimately limit production from this resource, even if sustained high oil prices make production profitable.

Nevertheless, the forecast is for the addition of significant tar sand capacity, which includes significant additions to capacity from both surface mining and in-situ production using injected steam (Pavone, 2009). Of the alternative methods, in-situ production is expected to grow at a much faster pace than conventional mining methods. In addition, North West Upgrading will install an upgrader in Sturgeon County, Canada, capable of upgrading 77 000 barrels per day of bitumen. The upgrader will use delayed coke (Lurgi) gasifiers and a hydrocracking unit utilizing the LC-Fining technology (Lummus Chevron). Commercial heavy oil developments have been considered (or are occurring) in Iran (Soroosh-Nowruz fields), Oman (Mukhaizna field) as well as Brazil (Marlim Sul and Jubarte) and Occidental Petroleum has taken a 15% share of Total's Joslyn project for $500 MM, replacing co-owner Enerplus Resources.

In terms of oil producing countries, the stability of the political situation in Canada is enviable and it is not surprising that Canadian oil sands production has seen repeated investment and has grown from relatively modest quantities in 1967 (Suncor, formerly Great Canadian oil Sands — 49,000 barrels per day of bitumen) and 1978 (Syncrude — 50,000 barrels per day of bitumen) to volumes in excess of 1 million barrels per day (TD Securities, 2007; Speight, 2007).

As a result, Athabasca the tar sands are making a significant contribution to Canada's total oil production, which is in excess of 3,000,000 million barrels per day (BP, 2008). Oil sand production will continue to increase and offset the decline in conventional crude oil production, ultimately becoming Canada's foremost source of oil - being projected to make up about two thirds of Canadian oil production during the next decade (http://www.mining-technology.com/projects/athabascasands/). At the time of writing, information obtained from literature sources that describe oil sand development indicate that production of synthetic crude oil from the Canadian oil sand deposits is on the order of 1,250,000 barrels per day. However, the type of tar sand reserves found at any given site dictates the extraction method that must be used (e.g., see Gingras and Rokosh, 2004; Speight, 2007) leading to variations in the cost of bitumen production.

Where tar sand deposits are relatively close to the surface, open-pit mining coupled with the hot water extraction process can be used to recover bitumen. Deeper deposits require in-situ methods such as steam injection through vertical or horizontal wells. This separates the bitumen from the sand in its underground reservoir prior to the liquid being pumped to the surface for collection and further processing.

The production of synthetic crude oil from the Canadian tar sands will help offset the overall North American decline in conventional light crude production and assist in meeting the expected increased demand for refined petroleum products. Refiners in the United States have already switched to running Canadian heavy crude oil, as well as blends of synthetic crude oil and bitumen (synbit) that compete against medium sour crude oil as refinery feedstock.

In June 2007, the potential impact of heavy oil and bitumen faced a setback when ExxonMobil Corporation and ConocoPhillips, two of the largest U.S. oil companies, abandoned their multi-billion dollar investments in the heavy oil and bitumen resources of the Orinoco basin in Venezuela (Pirog, 2007). This action followed the breakdown of negotiations between the companies and the government of President Hugo Chavez and Petróleos de Venezuela SA (PDV), the Venezuelan national oil company. However, four other international oil companies, including Total SA from France, Statoil from Norway, BP from Great Britain, and Chevron from the United States, accepted agreements that raised the PDV share in their Orinoco projects from approximately 40% to a controlling interest of about 78%.

However, producing synthetic crude oil from the Canadian tar sands is significantly more costly than producing conventional crude oil. Tax and royalty conditions in some jurisdictions often result in an economic environment for an operating company where its actual cost becomes far higher than simple production cost because of these taxes. As a result, Canadian tar sands economics, when compared to other tax and royalty environments, can be more desirable. Unfortunately, below a certain nominal price of oil, Canadian tar sands are not economically viable even if there are no royalty considerations. For existing tar sand producers with depreciated assets that were built when the cost of construction was far lower than it is today, that breakeven point is thought to be somewhere around $25 per barrel, whereas for new projects, the breakeven point may be closer to $100 per barrel (Pavon, 2009).

Synthetic crude oils derived from tar sands and other heavy hydrocarbons are priced at a significant discount to benchmark crude oils that are both lighter (lower viscosity) and sweeter (lower sulfur content). In order to upgrade synthetic crude oils from tar sands to the same yield of clean refined fuels, significantly more refinery processing is required.

Whether a refiner can be profitable or not processing tar sand-derived crude oil can be measured quantitatively using an industry standard called refinery crack spread. Although different organizations define the crack spread in similar ways, we present here refinery crack spread as defined by the New York Mercantile Exchange (NYMEX) and calculates the refinery crack spread as the difference between the sales price of a unit of refined fuels compared to the purchase price of the amount of light and sweet crude oil (benchmark crude oil) required to produce the refined fuels. The NYMEX assumes in its calculation that one barrel of crude oil produces one barrel of refined fuel. Although this is not precisely true, it serves as an easy means for calculating refinery crack spread.

Furthermore, approximately 10% of the petroleum produced worldwide is converted to petrochemicals, with most of the balance used for transportation fuels. The fundamental feedstocks used to produce petrochemicals from crude oil are refinery byproducts LPG and naphtha, and a smaller quantity of natural gas production byproduct natural gas liquids (NGLs). NGLs are primarily composed of ethane, propane, butane, pentane and some hexane. These fundamental petrochemical feedstocks are converted primarily to olefins, including ethylene, propylene, butadiene and butylene, and aromatics, including benzene, toluene and xylenes, through the processes of steam cracking and catalytic naphtha reforming to produce petrochemical derivatives (monomers and polymers).

The bitumen and synthetic crude oil form the Canadian tar sands also provide the potential for contributing some byproducts that could be used as petrochemical feedstocks. Bitumen upgrading processes, or delayed coking, gasification, hydrocracking, and hydroprocessing, produce the same starting materials for producing petrochemicals. Syngas produced from gasification could also produce hydrogen, ammonia, methanol, oxo-alcohols, Fischer-Tropsch liquids, and their derivatives.

In terms of pricing, tar sand bitumen is priced at a slight discount to Canadian heavy crude oil (less than $1 per barrel discount), while Canadian heavy crude oil is priced on average at a $10 per barrel discount to a mixture of conventional Canadian light and medium crude oils. However, the fragile nature of tar sands profitability was obvious during a period of rapidly falling crude oil price declines (Pavone, 2009). Total's oil sands project, named Joslyn, requires an oil price of just below $90 a barrel to achieve a 12.5% internal rate of return, while Total's developments in the deep waters off Angola need about $70 a barrel. Similarly, production of tar sand bitumen by in situ methods (mostly steam-assisted gravity drainage — SAGD — and cyclic steam stimulation – CSS) may meet economic restraints. It is, nevertheless, dependent upon the price the consumer is willing to pay for stability of supply to offset the fragile supply infrastructure that is often subject to political constraints and manipulation.

8.7 References

Alhajji, A.F., and Williams, J.L. 2003. Measures of Petroleum Dependence and Vulnerability in OECD Countries. Middle East Economic Survey. 46: 16. April 21.

Annual Energy Outlook. 2008. Annual Energy Outlook 2008 with Projections to 2030. Report No. DOE/EIA-0383(2008). Energy Information Administration, Department of Energy, Washington, DC. June.

BP. 2008. BP Statistical Review of World Energy. www.bp.com/statisticalreview.

Brandt, A. R. 2007. Testing Hubbert. Energy Policy, 35(5): 3074–3088.

Campbell Colin J (2004a). *The Essence of Oil & Gas Depletion*. Multi-Science Publishing, Brentwood, Essex, England.

Campbell Colin J (2004b). *The Coming Oil Crisis*. Multi-Science Publishing, Brentwood, Essex, England.

Campbell Colin J (2005). *Oil Crisis*. Multi-Science Publishing, Brentwood, Essex, England.

Campbell, C.J. and Laherrère, J.H. 1998. The End of Cheap Oil. Scientific American. 278: 78–83.

Cohen, D. 2007. The Perfect Storm. ASPO-USA/Energy Bulletin. October 31. http://www.energybulletin.net/node/36510

Deffeyes, K.S. 2002. Hubbert's Peak: The Impending World Oil Shortage. Princeton University Press, Princeton, New Jersey.

Deffeyes, K.S. 2005. Beyond Oil: The View from Hubbert's Peak. Hill and Wang, New York.

EIA. 2007. International Energy Outlook: Petroleum and Other Liquid Fuels. Energy Information Administration, Washington, DC.

Doukas, H., Flamos, A., and Psarras, J. 2008. Risks on Security of Oil & Gas Supply. Energy Sources, Part B: Economics, Planning and Policy. In press.

Eagles, L. 2006. Medium Term Oil Market Report. OECD/International Transport Forum Roundtable. International Energy Agency, Paris, France.

El-Genk, M.S. 2008. On the introduction of nuclear power in Middle East countries: Promise, strategies, vision and challenges. Energy Conversion and Management 49: 2618–2628.

Fisher W.H., and Kist O.F. 2001. Catastrophe Exposures and Insurance Industry Catastrophe Management Practices. American Academy of Actuaries Catastrophe Management Work Group.

Fleming, D. 2000. After Oil. Prospect Magazine, November. Issue No. 57: pp 12–13. http://www.prospect-magazine.co.uk.

Gerdes, J. 2007. Modest Non-OPEC Supply Growth Underpins $60+ Oil Price. SunTrust Robinson Humphrey, Atlanta, Georgia.

Gordon, 2007. Crude Oil: Uncertainty about Future Oil Supply Makes It Important to Develop a Strategy for Addressing a Peak and Decline in Oil Production. Report No. GAO-07-283. Report to Congressional Requesters. General Accountability Office, Washington, DC. February.

Jackson, P.M. 2007. Finding the Critical Numbers: What Are the Real Decline Rates for Global Oil Production? Cambridge Energy Research Associates, Cambridge, Massachusetts.

Hirsch, R.L., Bezdek, R., and Wendling, R. 23005. Peaking Of World Oil Production: Impacts, Mitigation, and Risk Management. February. http://www.netl.doe.gov/publications/others/pdf/Oil_Peaking_NETL.pdf.

Hubbert, M.K. 1956. Nuclear Energy and the Fossil Fuels - Drilling and Production Practice. Institute, Washington, DC.

Jackson, P.M. 2006. Why the Peak Oil Theory Falls Down: Myths, Legends, and the Future of Oil Resources. Cambridge Energy Research Associates, Cambridge, Massachusetts. November 14.

Laherrère, J.H. 2001. Paper presented at the EMF/IEA/IEW Meeting IIASA, Laxenburg, Austria. Plenary Session I: Resources, June 19.

NPC. 2007. Facing the Hard Truths about Energy – A Comprehensive View to 2030 of Global Oil and Natural Gas. National Petroleum Council, Washington, DC.

Pavone, T. 2009. Oil Sands Industry Status. Petroleum Technology Quarterly, Q1: 65–76.

Simmons M.R. 2000 Fighting Rising Demand and Rising Decline Curves: Can the challenge be met? SPE Asia Pacific Oil and Gas Conference, Yokohama, April 25. http://www.simmonscointl.com/web/downloads/spe.pdf

Söderbergh, B.; Robelius, F.; Aleklett, K. 2007. A Crash Program Scenario for The Canadian Oil Sands Industry. Energy Policy. 35(3): 1931–1947.

Williams, J.L., and Alhajji, A.F. 2003. The Coming Energy Crisis? Oil & Gas Journal. 101(5). February 3.

Yergin, Y. 1993. The Prize: The Epic Quest for Oil, Money and Power. Free Press, New York.

9

The Future

Conventional crude oil resources comprise all crude oil that is technically producible from reservoirs through a well bore using any primary, secondary, and enhanced recovery method (Chapter 1). This definition excludes liquids produced from mined deposits or created liquids such as liquids from natural gas. Furthermore, conventional crude oil is a large but finite volume insofar as all or very nearly all of major oil-bearing basins are identified, and production is clearly past its peak in some of the basins (Campbell and Laherrère, 1998).

Crude oil had influenced major changes in the trading and economic status of the United States over the past six decades (Table 9.1), and whether the status that existed in the 1950s can ever be rejuvenated is unlikely. The United States has come not only to rely on crude oil but the nation is also addicted to crude oil. Cures for this addition are possible, such as a reduction in the amount of oil required for daily life, but will take time and are unlikely to succeed in the near term. The misfortunes of the United States are rooted in the depletion of the domestic oil reserves — the resource that made the United States the most powerful nation in history.

Table 9.1 Change of economic status of the United States over the past six decades.

The U.S. in 1950	The U.S. in 2008
Self sufficient in oil: World's foremost oil producer World's foremost oil exporter	Lacks self sufficiency in oil World's foremost oil importer
Self-sufficient in nearly all resources	World's foremost importer of non-petroleum resources
World's major exporter of manufactured goods	World's largest importer of manufactured goods and manufacturing jobs outsourced to other countries
Major producer of premium automobiles	Auto-industry in financial trouble
World's foremost creditor nation	World's foremost debtor nation

Bad management by the government and by the private sector of the resources over the past six decades also adds to the equation and has contributed to the current situation.

As a result, oil depletion is real not only in the United States but also on a worldwide scale and most studies estimate that oil production will peak sometime between now and 2040. This range of estimates is wide because the timing of the peak depends on multiple, uncertain factors that will help determine how quickly the oil remaining in the ground is used, including the amount of oil still in the ground; how much of that oil can ultimately be produced given technological, cost, and environmental challenges as well as potentially unfavorable political and investment conditions in some countries where oil is located; and future global demand for oil (Gordon, 2007).

The world may never actually run out of crude oil only because oil will eventually become less expensive than lower cost alternative sources of fuels (Speight, 2008, 2009, 2011). The time frame of such development is dependent upon the realization that alternative fuels will become a way of life. However, alternative fuels and transportation technologies face challenges that could impede their ability to mitigate the consequences of a peak and decline in oil production, unless sufficient time and effort are brought to bear. For example, although corn ethanol production is technically feasible, it is more expensive to produce than gasoline and will require costly investments in infrastructure, such as pipelines and storage tanks, before it can become widely available as a primary fuel. Key alternative technologies currently supply the equivalent of only about 1% of U.S. consumption of petroleum products, and the Department of Energy (DOE) projects that even by 2015, they could displace only the equivalent of 4% of projected U.S. annual consumption. In such circumstances, an imminent peak and sharp decline in oil production could cause a worldwide recession.

However, if the peak is delayed, these technologies have a greater potential to mitigate the consequences. DOE predicts that these technologies could displace up to 34% of U.S. consumption in the 2025 through 2030 time frame, if the challenges are met. The level of effort dedicated to overcoming challenges will depend in part on sustained high oil prices to encourage sufficient investment in and demand for alternatives.

Depending on price assumptions, world unconventional production is projected to be 5.4 to 18.9 million barrels per day higher in 2030 than it was in 2006, accounting for between 6 and 22% of the world's total production of liquids (Annual Energy Outlook, 2008) and may even play a key pivotal supply role (Steinhubl *et al.*, 2009). Production of unconventional liquids depends heavily on prices, being more competitive with conventional sources when market prices are high. However, not all unconventional liquids respond to price changes in the same manner because the sources of unconventional liquids (Table 9.2) differ with regard to resource constraints, political backing, available technologies, and other characteristics.

Without ample low priced crude oil supplies, there is little realistic hope of restoring the world economy to the growth model that was desired prior to the run up in energy prices. In the United States, a world leading powerful economy was built on the back of cheap energy supplies, especially crude oil. Crude oil is the raw material input for so many products, such as gasoline, jet fuel, and plastics, that scarcity and high prices will lead to a complete transformation of our consumer driven business model. The United States is not nearly prepared for the transformation of the economy that will soon come. The age of cheap easily accessed crude oil supplies is nearly at an end. Even the current low price due to the worldwide deleveraging of debt is bad news for the long-term health of the

Table 9.2 Future sources of fuels.

- **Non-conventional oils**:
  - extra-heavy - Orinoco oil belt Venezuela
  - tar sands - Athabasca Canada (in-situ, mining)
  - synthetic crude (from tar, gas, coal)
- **Non-conventional gas:**
  - coalbed methane (CBM)
  - gas shale
  - tight gas sands
  - gas in geopressured aquifers
- **Conventional hydrocarbons in non-conventional locations**:
  - deepwater (>500 meters or <1000 meters)
  - Arctic
  - Antarctic
- **Uneconomic conventional hydrocarbons:**
  - very small fields: 1–10 Mb - (government take > large fields!)
  - special EOR (enhanced oil recovery)
  - HPHT (high pressure = 700 bars, high temperature = 150°C)
- **Surrogate non-conventional hydrocarbons:**
  - **oil shales** (immature source rock = not yet HC)
  - gas hydrates (oceans, solid, untrapped, dispersed = myth)
  - gas from the mantle = speculation

economy. The model of driving great distances from houses in the suburbs to businesses in the cities, with huge shopping malls in between, will soon be viewed as one of history's great mistakes of economic development.

Nevertheless, the demand for crude oil is largely influenced by global economic growth and will also be affected by government policies on the environment and climate change and consumer choices about conservation. Indeed, the precise status of crude oil as a continuing fuel source is unknown, speculative, and often subject to inspired guesswork.

In this supply constrained environment, other sources of liquid fuels need to be examined as it is certain that the demand for liquid fuels will increase into the foreseeable future and the potential for the use of sources energy as sources of fuel is worth comment at this time. More specifically, comments on the potential for undiscovered oil, coal, oil shale, and biomass, as well as the challenges for industry are warranted here.

9.1 Undiscovered Oil

Undiscovered oil resources are resources that are estimated to exist on the basis of broad geologic knowledge and theory, in undiscovered accumulations outside of known fields (Attanasi, 1999). To some, undiscovered oil is real while to others undiscovered oil is a myth.

Petroleum exploration is an efficient technical procedure. Investigating a geologic formation or a sedimentary basin using modern seismic techniques will reveal virtually all significant prospects, thus showing oil companies where to lease for further test drilling. Currently, there are virtually no areas where petroleum exploration cannot be successfully carried out if regional geological studies indicate a good chance of finding major petroleum fields (i.e. those with an ultimate recovery of more than 100 million barrels of oil).

Unfortunately, despite intense efforts by all of the world's oil companies, only a few of the new major fields promised by their geologists were actually found. The world's accessible oil provinces had all been previously recognized and most of their major fields found earlier. Numerous major finds had been made in the late 1960s, which brought on production offshore by new marine technologies during the mid-1970s in time to bring the OPEC producers to heel. No new major oil provinces that produce 7 to 25 billion barrels) have been found since 1980. The world is finite. The peak global oil finding year was 1962. Since then, the global discovery rate has dropped sharply in all regions and the 1,311 known major and giant oil fields contain 94% of the world's known oil and are accordingly the most critical for future global oil supplies.

Modern three-dimensional seismic and horizontal-drilling techniques improved current oil recovery in known fields, but made no substantial change in global reserves or discoveries of major fields. When the world oil price collapsed in 1986, exploration funds and efforts were cut back drastically everywhere; and by 1989, all major companies were downsizing and eliminating most of their geological and geophysical staffs. The minimum six-year period needed to discover the five largest fields in any basin had passed without making enough discoveries to whet top management's enthusiasm, so the money dried up for all but prime prospects. This is unfortunate because the huge remaining postulated resources will never be converted to actual reserves unless discoveries are made.

In addition, there is a great deal of disagreement on the issue of future oil supply; one reason is that there is confusion among the terms used, such as active and inactive reserves as well as known and unknown resources. In addition, there is no means by which foreign reserves can be accurately checked, leaving the numbers open to serious doubt (Campbell and Laherrère, J.H. 1998). Some of the reserve numbers may only be gross approximations.

Thus, it is essential to recognize that estimates of undiscovered oil and gas resources are just that — estimates and the reality of such estimates will, or should, always, be questioned. They are an attempt to quantify something that cannot be accurately known until the resource has been essentially depleted. For that reason, resource estimates should be viewed as assessed at a point in time based on whatever data, information and methodology were available at that time. Resource estimates therefore are subject to continuing revision as undiscovered resources are converted to reserves and as improvements in data and assessment methods occur (Attanasi, 1999).

Historically, estimates of the quantities of undiscovered oil and gas resources expected to exist within a region or the nation have been prepared for a variety of purposes using several different methods. To make effective use of such estimates, or to compare them with others, one must develop an understanding of how and why they were prepared; the extent and reliability of the data upon which they are based; the expertise of the assessors; the implications and limitations of the methodology used; and the nature of any geographic, economic, technologic, or time limitations and assumptions that may apply. It is equally important that those who prepare estimates provide documentation adequate to allow the users to evaluate the issues just described.

The purpose of this chapter is to examine, in general terms, some of these issues and how they may impact on the credibility and usefulness of resource estimates. In general, uncertainties in estimates of undiscovered oil and natural gas are greatest for areas that have had little or no past exploratory effort. For areas that have been extensively explored and are in a mature development stage, many of the unknowns have been eliminated and future resources can be evaluated with much more confidence. Even in some mature producing areas, however, uncertainty remains about the potential oil and gas supply at greater drilling depths. Uncertainty also pervades projections of whether potential reservoirs have been unrecognized or bypassed in past drilling. Similarly, where resource estimates are based on analogue comparisons between maturely explored areas and unexplored areas, uncertainty is introduced because each area or basin has unique characteristics.

Although our fundamental knowledge of the origin, migration, and entrapment of oil and gas has advanced markedly during the past 30 years, the fact that incremental scientific advances are still being made leads to additional uncertainty in resource estimation, especially in frontier areas or at great drilling depths. In other words, new knowledge may lead to increases or decreases in estimates of undiscovered resources, but generally leads to a reduction of uncertainty.

Discovery is only the first step in crude oil and natural gas resource development. The present state of technical knowledge in reservoir geology and petroleum engineering, as well as existing regulations, determine the spacing, completion, and production methods of development wells (i.e. petroleum producing wells drilled after the discovery well). As engineering and geologic knowledge increase, our ability to withdraw larger increments of oil and gas from existing fields is enhanced. Thus, the sizes of fields in terms of ultimately producible barrels of oil or cubic feet of gas increase with time. Uncertainty as to ultimate sizes of discovered fields leads to uncertainty in estimates of size distributions of undiscovered fields in areas with analogous reservoir characteristics and geologic histories.

Scientists can estimate the quantity of technically recoverable undiscovered oil and gas based on the present state of geological and engineering knowledge, modified by a consideration of future technological advancement. However, the percentage of that quantity that may actually be discovered and produced is an economic question. Uncertainties about future well-head crude oil and natural gas prices and costs of exploration and development adversely affect all resource estimates. In short, uncertainties embodied in economic assumptions lead to significant uncertainties in estimates of economically recoverable resources and account for some of the large differences among estimators.

9.2 Coal

Coal is a fossil fuel formed in swamp ecosystems where plant remains were saved by water and mud from oxidation and biodegradation. Coal is also a combustible organic sedimentary rock, composed primarily of carbon, hydrogen, and oxygen, formed from ancient vegetation and consolidated between other rock strata to form coal seams. The harder forms, such as anthracite coal, can be regarded as organic metamorphic rocks because of a higher degree of maturation.

Coal is composed primarily of carbon, along with assorted other elements, including sulfur. It is the largest single source of fuel for the generation of electricity worldwide, as well as the largest source of carbon dioxide emissions, which have been implicated as the primary cause of global warming. Coal is extracted from the ground by coal mining, either underground mining or open-pit mining (surface mining).

A fossil energy source, coal can play a substantial role as a transitional energy source as one moves from the petroleum and natural gas based economic system to the future economic system, based on non-depletable or renewable energy systems (BP, 2008; Speight, 2008). It has been used as an energy source for thousands of years and has many important uses, but most significantly in electricity generation, steel and cement manufacture, and industrial process heating.

Coal accounts for about 28% (hard coal 25%; soft brown coal 3%) of global primary energy consumption, surpassed only by crude oil (BGR, 2007). Developing countries use about 55% of the world's coal today; this share is expected to grow to 65% over the next 15 years (Balat and Ayar, 2004). In year 2050, coal will account for more than 34% of the world's primary energy demand.

In the developing world, the use of coal in the household for heating and cooking, is important (Balat, 2007). For coal to remain competitive with other sources of energy in the industrialized countries of the world, continuing technological improvements in all aspects of coal extraction have been necessary. Coal is often the only alternative when low-cost, cleaner energy sources are inadequate to meet growing energy demand (Balat and Ayar, 2004).

Coal occurs in different forms or types due to variations in the nature of the source material and local or regional the variations in the coalification processes cause the vegetal matter to evolve differently.

Thus, as geological processes increase their effect over time, the coal precursors are transformed over time into the following:

1. Lignite, also referred to as brown coal, is the lowest rank of coal and used almost exclusively as fuel for steam-electric power generation.
2. Sub-bituminous coal: the properties range from those of lignite to those of bituminous coal and are used primarily as fuel for steam-electric power generation.
3. Bituminous coal: a dense coal, usually black, sometimes dark brown, often with well-defined bands of bright and dull material, used primarily as fuel in steam-electric power generation, with substantial quantities also used for heat and power applications in manufacturing and to make coke.
4. Anthracite: the highest rank; a harder, glossy, black coal used primarily for residential and commercial space heating.

The rank of a coal indicates the progressive changes in carbon, volatile matter, and probably ash and sulfur that take place as coalification progresses from the lower-rank lignite through the higher ranks of sub-bituminous, high-volatile bituminous, low-volatile bituminous, and anthracite. The rank of a coal should not be confused with its grade. A high rank (e.g., anthracite) represents coal from a deposit that has undergone the greatest degree of metamorphosis and contains very little mineral matter, ash, and moisture. On the other hand, any rank of coal, when cleaned of impurities through coal preparation will be of a higher grade.

The production of liquid fuels from coal is not new and has received considerable attention because the concept does represent alternate pathways to liquid fuels (Berthelot, 1869; Batchelder, 1962; Anderson and Tillman, 1979; Whitehurst *et al.*, 1980; Gorin, 1981; Stranges, 1983; Stranges, 1987; Argonne, 1990; Speight, 2008). The first major use of Fischer Tropsch technology was during the Second World War, when Germany produced about 90% of the diesel and aviation fuel. South Africa began liquefying coal in response to apartheid-era sanctions, and in part as a result of its investment back then, continues to derive about 30% of its fuel from liquefied coal. In fact, the concept is often cited as a viable option for alleviating projected shortages of liquid fuels as well as offering some measure of energy independence or energy security for those countries with vast resources of coal who are also net importers of crude oil.

One of the early processes for the production of liquid fuels from coal involved the Bergius process. In the process, lignite or sub-bituminous coal is finely ground and mixed with heavy oil recycled from the process. Catalyst is typically added to the mixture and the mixture is pumped into a reactor. The reaction occurs at between 400 to 500°C and 20 to 70 MPa hydrogen pressure. The reaction produces heavy oil, middle oil, gasoline, and gas:

$$nC_{coal} + (n+1)H_2 \rightarrow C_nH_{2n+2}$$

For the conversion of coal to liquids, referred to as direct coal liquefaction, the product of synthetic fuel depends on the coal type, the reactor configuration, the process parameters, and the catalysts. In addition, the different product fractions need further processing to yield synthetic fuel or a fuel blending stock of the desired quality.

More recently, other processes have been developed for the conversion of coal to liquid fuels. The Fischer-Tropsch process of indirect synthesis of liquid hydrocarbons and is currently today used by Sasol in South Africa. In the process, referred to as indirect coal liquefaction, because the coal is first converted to gas, coal is gasified to make syngas — a balanced purified mixture of CO and H_2 gas — and the syngas condensed using Fischer-Tropsch catalysts to make light hydrocarbons which are further processed into gasoline and diesel. Syngas can also be converted to methanol.

In spite of the interest in coal liquefaction processes that emerged during the 1970s and the 1980s, petroleum prices always remained sufficiently low to ensure that the initiation of a synthetic fuels industry based on non-petroleum sources would not become a commercial reality.

With the continuing uncertainty among the oil-producing countries and the violence in the Middle East, the issue of energy security is again of prime importance. Part of this energy security can come by the production of liquid fuels from coal. As always, the real question relates to economics, tied to the price of oil.

The Sasol technology, a third-generation Fischer-Tropsch process, was developed in Germany and used in World War II, and later in South Africa. In the process, steam and oxygen are passed over coke at high temperatures and pressures. Synthesis gas (a mixture of carbon monoxide and hydrogen) is produced and then reassembled into liquid fuels through the agency of the Fischer Tropsch reaction (Speight, 2008).

Although considered too expensive to compete with the production of liquid fuels from petroleum, the liquid fuels produced from coal by gasification and the conversion of the synthesis gas have environmentally friendly insofar as they have no sulfur and, therefore, are more in keeping with the recent laws regarding ultra-low sulfur fuels.

In fact, the technologies required to produce large-scale supplies of clean liquid fuels from coal are not on the drawing boards or in laboratories. They are in use around the world today, from countries including South Africa, which has long relied on coal liquefaction to provide a substantial percentage of its transportation fuels, to China, India, Indonesia and the Philippines.

The obvious drawback to direct and indirect processes that produce liquids from coal relates to environmental issues. The process of converting coal into liquid and using it for transportation releases nearly twice as much carbon dioxide as burning diesel made from crude oil does. In a world conscious of climate change, that excess carbon is a major issue. One way around this problem might be to take the carbon dioxide and bury it underground. Another would be to replace fossil-fuel feedstock with biomass (Speight, 2008).

Economics is another issue since the processes (especially the Fischer-Tropsch technology) has always been relatively expensive. However, the need may have to justify the expense insofar as energy independence is the matter at hand.

9.3 Oil Shale

Oil shale is a complex mixture of organic and inorganic materials that vary widely in composition and properties. In general terms, oil shale is a fine-grained sedimentary rock that is rich inorganic matter and yields oil when heated. Some oil shale is genuine shale but others have been incorrectly classified and are actually siltstones, impure limestone, or even impure coal.

The oil shale deposits in the western United States contain approximately 15 percent organic material, by weight. In the United States, the deposits are concentrated in the Green River Formation in the states of Colorado, Wyoming and Utah, account for nearly three-quarters of this potential. The amount of kerogen in the shale varies with depth, with the richer portions appearing much darker. For example, in Colorado, the richest layers are termed the Mahogany Zone after the rich brown color.

Oil shale does not contain oil and only produces oil when it is heated to about 500°C (about 932°F), when some of the organic material is transformed into a distillate. The potential for liquids production potential from oil shale is measured by a laboratory pyrolysis method called Fischer Assay (Scouten, 1990; Speight, 1994) and is reported in barrels (42 gal) per ton. Rich zones can yield more than 40 gallons per ton, while most shale falls in the range of 10 to 25 gallons per ton. Oil shale yields more than 25 United States gal/ton are generally viewed as the most economically attractive, and hence, the most favorable for initial development. Shale oil can be refined into synthetic crude oil and the *known* resources of the United States could translate into as much as 2.2 trillion (2.2×10^{12}) barrels of synthetic crude oil.

Surface processes for the production of liquids involve *mining* the shale followed by heating at the surface in a reactor, referred to as retorting, produces shale oil, which is often referred to referred to as synthetic crude oil both before and after hydrotreating. However, with the exception of in situ processes, oil shale must be mined before it can be converted to shale oil. Depending on the depth and other characteristics of the target oil shale deposits, either surface mining or underground mining methods may be used.

In-situ processes avoid the need to mine the shale but introduce heat to the kerogen while it is still embedded in its natural geological formation and the product is recovered as distillate through a well. There are two general in-situ approaches; true in-situ in which there is minimal or no disturbance of the ore bed, and modified in-situ, in which the bed is given a rubble-like texture, either through direct blasting with surface up-lift or after partial mining to create void space. In situ processes tend to operate slowly and may require one or more years to produce shale oil.

Modified in-situ processes involve mining beneath the target oil shale deposit prior to heating. Such processes attempt to improve performance by exposing more of the deposit to the heat source and by improving the flow of gases and liquids through the rock formation, and increasing the volumes and quality of the oil produced.

Shale retorting processes produce oil with almost no high-boiling residuum. With upgrading, shale oil is a light boiling premium product more valuable than most crude oils. However, the properties of shale oil vary as a function of the production (retorting) process. Fine mineral matter carried over from the retorting process and the high viscosity and instability of shale oil produced by present retorting processes have necessitated upgrading of the shale oil before transport to a refinery.

Upgrading, or partial refining, to improve the properties of a crude shale oil may be carried out using different options. Hydrotreating is the option of choice to produce a stable product that is comparable to benchmark crude oils (Chapter 4). In terms of refining and catalyst activity, the nitrogen content of shale oil is a disadvantage. If not removed, the arsenic and iron in shale oil would poison and foul the supported catalysts used in hydrotreating.

Blending shale oil products with corresponding crude oil products, using shale oil fractions obtained from a very mildly hydrogen treated shale oil, yields kerosene and diesel fuel of satisfactory properties. Hydroprocessing shale oil products, either alone or in a blend with the corresponding crude oil fractions, is therefore necessary. The severity of the hydroprocessing has to be adjusted according to the particular properties of the feed and the required level of the stability of the product.

The fundamental problem with all oil shale technologies is the need to provide large amounts of heat energy to decompose the kerogen to liquid and gas products. More than one ton of shale must be heated to temperatures in the range 850° and 1000°F (425 to 525°C) for each barrel of oil generated, and the heat supplied must be of relatively high quality to reach retorting temperature. Once the reaction is complete, recovering sensible heat from the hot rock is very desirable for optimum process economics. This leads to three areas where new technology could improve the economics of oil recovery:

1. Recovering heat from the spent shale.
2. Disposal of spent shale, especially if the shale is discharged at temperatures where the char can catch fire in the air.
3. Concurrent generation of large volumes of carbon dioxide.

The heat recovery from hot solids is generally not efficient, unless it is in the area of fluidized bed technology. However, to apply fluidized bed technology to oil shale would require grinding the shale to sizes less than about one millimeter, an energy intensive task that would result in an expensive disposal problem. Such fine particles might be used in a lower temperature process for sequestering carbon dioxide (Fenton, 1977).

Disposal of spent shale is also a problem that must be solved in economic fashion for the large-scale development of oil shale to proceed.

Retorted shale contains carbon as char, representing more than half of the original carbon values in the shale. The char is potentially pyrophoric and can burn if dumped into the open air while hot. The heating process results in a solid that occupies more volume than the fresh shale because of the problems of packing random particles. A shale oil industry producing 100,000 barrels per day — about the minimum for a world-scale operation — would process more than 100,000 tons of shale and result in an amount of spent shale that is equivalent to a block more than 100 feet on a side, assuming some effort at packing to conserve volume. Part of the spent shale could be returned to the mined-out areas for remediation, and some can potentially be used as feed for cement kilns.

In situ processes avoid the spent shale disposal problems because the spent shale remains where it is created but, on the other hand, the spent shale will contain uncollected liquids that can leach into ground water, and vapors produced during retorting can potentially escape to the aquifer (Karanikas *et al.*, 2005).

As the demand for light hydrocarbon fractions constantly increases, there is much interest in developing economical methods for recovering liquid hydrocarbons from oil shale on a commercial scale. However, the recovered hydrocarbons from oil shale are not yet economically competitive against the petroleum crude produced. Furthermore, the value of hydrocarbons recovered from oil shale is diminished because of the presence of undesirable contaminants. The major contaminants are sulfurous, nitrogenous, metallic, and organometallic compounds, which cause detrimental effects to various catalysts used in the subsequent refining processes. These contaminants are also undesirable because of their disagreeable odor, corrosive characteristics, and combustion products that further cause environmental problems.

Oil shale still has a future and remains a viable option for the production of liquid fuels. Many of the companies involved in earlier oil shale projects still hold their oil shale technology and resource assets. The body of knowledge and understanding established by these past efforts provides the foundation for ongoing advances in shale oil production, mining, retorting, and processing technology and supports the growing worldwide interest and activity in oil shale development. In fact, in many cases, the technologies developed to produce and process kerogen oil from shale have not been abandoned, but rather mothballed for adaptation and application at a future date when market demand would increase and major capital investments for oil shale projects could be justified.

In terms of innovative technologies, both conventional and in-situ retorting processes result in inefficiencies that reduce the volume and quality of the produced shale oil. Depending on the efficiency of the process, a portion of the kerogen that does not yield liquid is either deposited as coke on the host mineral matter, or is converted to hydrocarbon gases. For the purpose of producing shale oil, the optimal process is one that minimizes the regressive thermal and chemical reactions that form coke and hydrocarbon gases and maximizes the production of shale oil. Novel and advanced retorting and upgrading processes seek to modify the processing chemistry to improve recovery and/or create high value byproducts. Novel processes are being researched and tested in lab-scale environments. Some of these approaches include lower heating temperatures, higher heating rates, shorter residence time durations, introducing scavengers, such as hydrogen or hydrogen transfer/donor agents, as well as introducing solvents (Baldwin, 2002).

Finally, the development of western oil shale resources will require water for plant operations, supporting infrastructure, and the associated economic growth in the region. While some oil shale technologies may require reduced process water requirements, stable and secure sources of significant volumes of water may still be required for large-scale oil shale development. The largest demands for water are expected to be for land reclamation and to support the population and economic growth associated with oil shale activity.

Nevertheless, if a technology can be developed to economically recover oil from oil shale and for meeting energy demand in an environmentally acceptable manner, the potential for oil shale is enormous (Bartis *et al.*, 2005; Andrews, 2006). If the kerogen could be converted to oil, the quantities would be far beyond all known conventional oil reserves. Unfortunately, the prospects for oil shale development are uncertain. The estimated cost of surface retorting remains high and many consider it unwise to move towards near-term commercial efforts.

9.4 Liquids from Biomass

Fossil fuels are finite energy resources (Salameh, 2005; Pimentel and Pimentel, 2006). Therefore, reducing national dependence of any country on imported crude oil is of critical importance for long-term security and continued economic growth. Supplementing petroleum consumption with renewable biomass resources is a first step towards this goal. The realignment of the chemical industry from one of petrochemical refining to a bio-refinery concept is, given time, feasible and has become a national goal of many oil-importing countries. However, clearly defined goals are necessary for increasing the use of biomass-derived feedstocks in industrial chemical production and it is important to keep the goal in perspective.

Biomass currently supplies 14% of the world's energy needs, but has the theoretical potential to supply 100%. Most present day production and use of biomass for energy is carried out in a very unsustainable manner with a great many negative environmental consequences. If biomass is to supply a greater proportion of the world's energy needs in the future, the challenge will be to produce biomass and to convert and use it without harming the natural environment. Technologies and processes exist today which, if used properly, make biomass based fuels less harmful to the environment than fossil fuels. Applying these technologies and processes on a site specific basis in order to minimize negative environmental impacts is a prerequisite for sustainable use of biomass energy in the future.

In this context, the increased use of biofuels should be viewed as one of a range of possible measures for achieving self-sufficiency in energy, rather than a panacea (Crocker and Crofcheck, 2006; Worldwatch Institute, 2006; Freeman, 2007; Nersesian, 2007).

Biomass is a renewable resource, whose utilization has received great attention due to environmental considerations and the increasing demands of energy worldwide (Tsai *et al.*, 2007; Speight, 2008). Biomass is clean for it has negligible content of sulfur, nitrogen and ash-forming constituents, which give lower emissions of sulfur dioxide, nitrogen oxides, and soot than conventional fossil fuels. The main biomass resources include the following: forest and mill residues, agricultural crops and wastes, wood and wood wastes, animal wastes, livestock operation residues, aquatic plants, fast-growing trees and plants, and municipal and industrial wastes.

The role of wood and forestry residues in terms of energy production is as old as fire itself and in many societies wood is still the major source of energy. In general, biomass can include anything that is not a fossil fuel that is bioorganic-based (Lucia *et al.*, 2006).

There are many types of biomass resources that can be used and replaced without irreversibly depleting reserves and the use of biomass will continue to grow in importance as replacements for fossil materials used as fuels and as feedstocks for a range of products (Narayan, 2007; Speight, 2008). Some biomass materials also have particular unique and beneficial properties that can be exploited in a range of products including pharmaceuticals and certain lubricants.

Biomass represents a potentially large economic resource is effectively being thrown away. For example, the straw associated with the wheat crop in often ploughed back into the soil, even though only a small proportion is needed to maintain the level of organic matter. Thus, a huge renewable resource is not being usefully exploited since wheat straw contains a range of potentially useful chemicals. These include the following:

1. Cellulose and related compounds that can be used for the production of paper and/or bioethanol
2. Silica compounds which can be used as filter materials, such as those necessary for water purification
3. Long-chain lipids that can be used in cosmetics or for other specialty chemicals.

For example, primary biomass feedstocks are thus primary biomass that is harvested or collected from the field or forest where it is grown. Examples of primary biomass feedstocks currently being used for bioenergy include grains and oilseed crops used for transportation fuel production, plus some crop residues, such as orchard trimmings and nut hulls, and some residues from logging and forest operations that are currently used for heat and power production. In the future, it is anticipated that a larger proportion of the residues inherently generated from food crop harvesting, as well as a larger proportion of the residues generated from ongoing logging and forest operations, will be used for bioenergy (Smith, 2006). Additionally, as the bioenergy industry develops, both woody and herbaceous perennial crops will be planted and harvested specifically for bioenergy and product end uses.

Secondary biomass feedstocks differ from primary biomass feedstocks in that the secondary feedstocks are a byproduct of processing of the primary feedstocks. By processing it is meant that there is substantial physical or chemical breakdown of the primary biomass and production of byproducts; processors may be factories or animals. Field processes such as harvesting, bundling, chipping or pressing do not cause a biomass resource that was produced by photosynthesis (e.g., tree tops and limbs) to be classified as secondary biomass.

Specific examples of secondary biomass include sawdust from sawmills, black liquor (which is a byproduct of paper making), and cheese whey (which is a byproduct of cheese making processes). Manures from concentrated animal feeding operations are collectable secondary biomass resources. Vegetable oils used for biodiesel that are derived directly from the processing of oilseeds for various uses are also a secondary biomass resource (Wright *et al.*, 2006; Bourne, 2007).

Tertiary biomass *feedstock* includes post consumer residues and wastes, such as fats, greases, oils, construction and demolition wood debris, other waste wood from the urban environments, as well as packaging wastes, municipal solid wastes, and landfill gases. *Other wood waste from the urban environment* includes trimmings from urban trees, which technically fit the definition of primary biomass. However, because this material is normally handled as a waste stream along with other post-consumer wastes from urban environments, it makes the most sense to consider it to be part of the tertiary biomass stream.

Biomass can produce a range of liquid fuels that exhibit a wide range of physical and chemical properties.

Alcohols are oxygenate fuels produced from biomass and practically any of the organic molecules of the alcohol family can be used as a fuel (Speight, 2008). The alcohols that can be used for motor fuels are methanol (CH_3OH), ethanol (C_2H_5OH), propanol (C_3H_7OH), butanol (C_4H_9OH). However, only methanol and ethanol fuels are technically and economically suitable for internal combustion engines (Bala, 2005).

Currently the production of ethanol (sometimes referred to as bio-ethanol) by fermentation of corn-derived carbohydrates is the main technology used to produce liquid fuels from biomass resources (McNeil Technologies Inc., 2005). Furthermore, amongst different biofuels, suitable for application in transport, bioethanol and biodiesel seem to be the most feasible ones at present. The key advantage of bioethanol and biodiesel is that they can be mixed with conventional petrol and diesel respectively, which allows using the same handling and distribution infrastructure. Another important strong point of bioethanol and biodiesel is that when they are mixed at low concentrations (≤10% bioethanol in petrol and ≤20% biodiesel in diesel, no engine modifications are necessary.

Ethanol can be blended with gasoline to create E85, a blend of 85% ethanol and 15% gasoline. E85 and blends with even higher concentrations of ethanol, E95, include pure bioethanol (E100-fuel), which has been used mainly in Brazil (Davis, 2006; Minteer, 2006; Speight, 2008). More widespread practice has been to add up to 20% to gasoline (E20-fuel or gasohol) to avoid engine changes. However, E100-fueled vehicles have difficulty starting in cold weather, but this is not a problem for E85 vehicles because of the presence of gasoline.

In comparison to gasoline, ethanol contains 35% oxygen by weight; gasoline contains none. The presence of the oxygen promotes more complete combustion, which results in fewer tailpipe emissions. Compared to the combustion of gasoline, the combustion of ethanol substantially reduces the emission of carbon monoxide, volatile organic compounds, particulate matter and green house gases. However, a unit of ethanol contains about 32% less energy than a unit of gasoline. One of the best qualities of ethanol is its octane rating.

Biodiesel is the generic name for fuels obtained by esterification (by methanol or by ethanol) of vegetable oil (Knothe *et al.*, 2005; Bockey, 2006). Biodiesel can be used in a diesel engine without modification and is a clean burning alternative fuel produced from domestic, renewable resources. The fuel is a mixture of fatty acid alkyl esters made from vegetable oils, animal fats or recycled greases. Where available, biodiesel can be used in compression-ignition (diesel) engines in its pure form with little or no modifications.

Biodiesel is biodegradable, nontoxic, and essentially free of sulfur and aromatics. It is usually used as a petroleum diesel additive to reduce levels of particulates, carbon monoxide, hydrocarbons, and air toxics from diesel-powered vehicles. When used as an additive, the resulting diesel fuel may be called B5, B10 or B20, representing the percentage of the biodiesel that is blended with petroleum diesel.

Biodiesel is produced through a process in which organic oils, such as soybean oil or recycled cooking oil, are combined with alcohol (ethanol or methanol) in the presence of a catalyst to form the ethyl or methyl ester. The biomass-derived ethyl or methyl esters can be blended with conventional diesel fuel or used as a neat fuel (100% biodiesel). Biodiesel can be made from any vegetable oil, animal fats, waste vegetable oils, or microalgae oils. Soybeans and Canola (rapeseed) oils are the most common vegetable oils used today.

Fuel-grade biodiesel must be produced to strict industry specifications (ASTM D6751) in order to insure proper performance. Biodiesel is the only alternative fuel to have fully completed the health effects testing requirements of the 1990 Clean Air Act Amendments and biodiesel that meets ASTM D6751 and is legally registered with the Environmental Protection Agency is a legal motor fuel for sale and distribution. Raw vegetable oil cannot meet biodiesel fuel specifications; therefore, it is not registered with the EPA and it is not a legal motor fuel.

A totally different process than that used for biodiesel production can be used to convert biomass into a type of fuel similar to diesel, which is known as bio-oil. The process *of fast pyrolysis or flash pyrolysis* occurs when solid fuels are heated at temperatures between 350 and 500°C for a very short period of time (<2 seconds. The bio-oils currently produced are suitable for use in boilers for electricity generation.

In another process, the feedstock is fed into a fluidized bed at 450 to 500°C and the feedstock flashes and vaporizes. The resulting vapors pass into a cyclone where solid particles, char, are extracted. The gas from the cyclone enters a quench tower where they are quickly cooled by heat transfer using bio-oil already made in the process. The bio-oil condenses into a product receiver and any non-condensable gases are returned to the reactor to maintain process heating. The entire reaction from injection to quenching takes only two seconds.

9.5 Energy Independence

Energy independence has been a political non-issue in the United States since the first Arab oil embargo in 1973. Since that time, the speeches of various presidents and the Congress of the United States have continued to call for an end to the dependence on foreign oil by the United States. Nevertheless, the United States has grown more dependent on foreign oil with no end in sight. For example, in 1970 the United States imported approximately one third of the daily oil requirement. Currently, the amount of imported oil is two thirds of the daily requirement. The congressional rhetoric of energy independence continues but meaningful suggestions of how to address this issue remain few and far between. The economy of the United States feeds on oil and the country consumes far more oil than it can produce.

Generally, the concept of energy independence for the United States runs contrary to the trend of the internationalization of trade. The government of the United States continues to reduce trade barriers through policies such as the North American Free Trade Agreement (NAFTA), the elimination of tariffs, as well as other free trade agreements. As a result, the percentage of the U.S. economy that comes from international trade is steadily rising.

Increased world trade is beneficial both economically and politically insofar as it is supposed to help establish amicable relations between countries. Through mutually beneficial exchange, a great deal of this increased interrelationship will, in theory, establish opportunities for personal ties that make military war or other forms of military action less likely. However, there are also contrary cases where countries acquire the means to be more destructive, if they so choose, through expanded economic opportunities. This type of argument has been used with regard to Iran.

Economic interdependence also makes the domestic economy more susceptible to disruptions in distant and unstable regions of the globe, such as the Middle East, South America, and Africa. In fact, in many countries with proven reserves, oil production could be shut down by wars, strikes, and other political events, thus reducing the flow of oil to the world market. If these events occurred repeatedly, or in many different locations, they could constrain exploration and production, resulting in a peak despite the existence of proven oil reserves. Using a measure of political risk that assesses the likelihood that events such as civil wars, coups, and labor strikes will occur in a magnitude sufficient to reduce a country's gross domestic product (GDP) growth rate over the next 5 years, four countries (Iran, Iraq, Nigeria, and Venezuela) possess proven oil reserves greater than 10 billion barrels and which countries contain almost one-third of worldwide oil reserves, face high levels of political risk. In fact, countries with medium or high levels of political risk contain 63% of proven worldwide oil reserves.

For example, in past years, disputes leading to withdrawal of labor by workers in Venezuela have caused reductions in the crude oil of approximately 1,500,000 barrels per day imported from that country. Similarly, conflicts in Nigeria between ethnic groups can disrupt the amount of crude oil of approximately 570,000 barrels per day imported from that country. Thus, in a short time and by events out of its control, the United States can suffer an oil shortage of imported oil to the tune of 2,000,000 barrels per day that leaves a large gap in the required 18,000,000 barrels per day currently refined in the United States.

Furthermore, the oil industry itself has been a story of vast swings between periods of overproduction, when low prices and profits led oil producers to devise ways to restrict output and raise prices, and periods when oil supplies appeared to be on the brink of exhaustion, stimulating a global search for new supplies. This cycle may now be approaching an end. It appears that world oil supplies may truly be reaching their natural limits. With proven world oil reserves anticipated to last less than 40 years, the age of oil that began near Titusville may be coming to an end. In the years to come, the search for new sources of oil will be transformed into a quest for entirely new sources of energy.

Political and investment risk factors continue to affect future oil exploration and production and, ultimately, the timing of peak oil production. These factors include changing political conditions and investment climates in many countries that have large proven oil reserves, which are important in affecting future oil exploration and production.

Even in the United States, political considerations may affect the rate of exploration and production. For example, restrictions imposed to protect environmental assets mean that some oil may not be produced. The Minerals Management Service of the United States Department of the Interior estimates that approximately 76 billion barrels of oil lie in undiscovered fields offshore in the outer continental shelf of the United States, which is necessary for a measure of energy security. Nevertheless, Congress enacted moratoriums on drilling and exploration in this area to protect coastlines from unintended oil spills. In addition, policies on federal land use need to take into account multiple uses of the land including environmental protection. Environmental restrictions may affect a peak in oil production by barring oil exploration and production in environmentally sensitive areas. The government must adopt policies that ensure our energy independence. The United States Congress is no longer believable when the members of the Congress lay the blame on foreign governments or events for an impending crisis. The Congress needs to look north and the positive role played by the Government of Canada in the early 1960s when the decision was made to encourage development of the Alberta tar sands. Synthetic crude oil production in no in excess of 1,000,000 million barrels per day — less than 6% of the daily liquid fuels requirement in the United States but a much higher percentage of Canadian daily liquid fuels requirement. In the United States, the issue to be faced is not so much oil reserves but oil policies.

The economics of crude oil inventories provides the key to unlocking this mystery. The net cost of carrying inventories is equal to the interest rate, plus the cost of physical storage, minus the convenience yield. The convenience yield is driven by the precautionary demand for the storage. When the convenience yield is zero, a market is in full carry, future prices exceed spot prices and inventories are abundant. Alternatively, when the precautionary demand for oil is high, spot prices are strong and exceed future prices, and inventories are unusually low.

The strategic reserve should be used to give the country a measure of energy independence and to thwart the efforts of the cartel to control oil.

Various importing countries are dependent on petroleum (Alhajji and Williams, 2003). In the United States, increasing petroleum imports is considered a threat to national security, but there is also the line of thinking that the level of imports has no significant impact on energy security, or even national security. However, the issue becomes a problem when import vulnerability increases as petroleum imports rise, which occurs when oil-consuming countries increase the share of petroleum imports from politically unstable areas of the world.

More generally, there are four measures of petroleum dependence (Alhajji and Williams, 2003):

1. Petroleum imports as a percentage of total petroleum consumption
2. The number of days total petroleum stocks cover petroleum imports
3. The number of days total stocks cover consumption
4. The percentage of petroleum in total energy consumption.

The dependency on foreign oil in the United States has increased steadily since 1986 and has reached record highs in the past decade two years — the degree of import dependence as percentage of consumption increased from about 50% in the early-to-mid 1980s to 60% in the early 1990s because of higher economic growth and lower oil prices on the demand side and declining US production on the supply side. Currently, 65–70% of the daily oil and oil products is now imported into the United States. There have been some minor fluctuations but the changes in the amount of oil imported into the United States are usually related to changes in the US economy and the US oil production.

The United States is the only oil-importing country with significant production, which is also a net importer. It is also unique in that among the net importers it has the lowest dependence in terms of net imports as a percent of consumption but it also has the highest absolute level of imports. The geo-political and economic interests and commitments raise the level of concern by US policy makers about dependence on imported oil.

In addition, commercial oil stocks in the United States have been at their lowest level three decades. Total petroleum inventories, which include commercial and stocks in the Strategic Petroleum Reserve are relatively low, in terms of daily coverage. Current commercial inventories are near the level at which spot shortages can occur. The past decade has seen scenarios in which the decline in commercial stocks is greater than the increase in the Strategic Petroleum Reserve and the capacity of the Strategic Petroleum Reserve and commercial stocks to deal with a crisis is less than before the refilling program began (Williams and Alhajji, 2003). Moreover, the premature release of petroleum from the Strategic Petroleum Reserve can jeopardize national security in case of continued political problems in the oil producing countries and weakens the ability of the United States to respond to real shortages.

Although some of the oil-importing countries have made progress in reducing their dependence on oil, the dependence of the United States on petroleum has increased in recent years from 38% of total energy consumption in 1995 to approximately 40% at the current time. This indicates two possible areas of concern regarding the extent to which petroleum influences energy security: the increase in the petroleum share of energy use, and the inability or unwillingness of the United States to reduce dependence on oil.

It might be argued that the degree of dependence has no impact on energy security as long as foreign oil is imported form secure sources. However, if the degree of dependence on non-secure sources increases, energy security would be in jeopardy. In this case, vulnerability would increase and economic and national security of individual oil-importing countries would be compromised.

One or more key suppliers can use the percentage of imports from the top five suppliers as a measure of the supply vulnerability to an interruption. This is an important measure of the vulnerability of the United States to supply disruption because it shows the high level of import concentration by importing from few suppliers. The top five suppliers to oil-importing countries are Saudi Arabia, the former USSR, Norway, Venezuela, and Mexico. The United States has a unique arrangement as well as location with Canada and Mexico but the political stability of Saudi Arabia is always open to discussion and question. In addition, problems in Venezuela underline the United States vulnerability to interruption from a major supplier. Venezuela has had severe problems and the potential for an interruption in oil supply always exist. Iraq crude has off the market for several years (oil supply from is only just starting again) and conflict in Nigeria has significantly influenced oil output.

Another important measure of vulnerability is the share of world crude coming from the Gulf region. The Gulf region has been viewed historically as a politically unstable area. Incidents in the region led to the three energy crises in 1973, 1979, and the two recent Gulf Wars. While a smaller share of imported oil from the Gulf producers means lesser vulnerability, it may increase vulnerability buy having to rely on other sources, such as Venezuela.

Whichever measure is used to assess energy dependence, the United States remains susceptible to an energy crisis because of the high dependence on imported oil. The United States has not made a significant reduction in dependence on imports since the mid 1980s and they continue to import almost 70% of the total petroleum consumption.

The petroleum share in the total energy supply also reflects the dependence on petroleum by the United States. In recent years; however, this share has increased in the United States. Because of this, the United States is highly vulnerable to oil supply disruptions.

Indeed, the possibility of energy crisis in the foreseeable future is greater than in previous years. Furthermore, the use of the Strategic

Petroleum Reserve, or government controlled stocks, to lessen the impact of an energy crisis is subject to debate. In fact, the premature release of oil stocks from the Strategic Petroleum Reserve may exacerbate an energy crisis as it depletes the stocks while shortages still exist since it can lead to stabilized or even lower prices and increased consumption (Alhajji and Williams, 2003).

Dependency and vulnerability to oil imports in the United States and, for that matter, in other oil-importing countries can be reduced by diversification of suppliers and by energy diversification. In addition, diversification of suppliers has the potential to lower the relative impact of supply disruption on most countries. The political instability that swings back and forth in countries such Venezuela, Nigeria, and Iraq emphasizes the need for diversification of suppliers and so removing the reliance on a small number of oil-producing countries.

9.6 Energy Security

Energy security is the continuous and uninterrupted availability of energy, to a specific country or region. The security of energy supply conducts a crucial role in decisions that are related to the formulation of energy policy strategies. The economies of many countries are depended by the energy imports in the notion that their balance of payments is affected by the magnitude of the vulnerability that the countries have in crude oil.

The Hubbert theory of peak oil (Chapter 8) assumes that oil reserves will not be replenished (i.e. that abiogenic replenishment is negligible) and predicts that future world oil production must inevitably reach a peak and then decline as these reserves are exhausted. Controversy surrounds the theory since as predictions for the time of the global peak is dependent on the past production and discovery data used in the calculation.

For the United States, the prediction turned out to be correct and, after the US peaked in 1971 and thus lost its excess production capacity, OPEC was able to manipulate oil prices. Since then, oil production in several other countries has also peaked. However, for a variety of reasons, it is difficult to predict the oil peak in any given region. Based on available production data, proponents have previously and incorrectly predicted the peak for the world to be in years 1989, 1995, or in the 1995 to 2000 period. Other predictions have chosen 2007 and beyond for the peak of oil production.

But more important, several trends that should have been established in the wake of the decreasing crude prices have never been put into practice. For example, and most important, the failure of politicians to recognize the need for a measure of energy independence through the development of alternate resources as well the development of technologies that would assists in maximizing oil recovery.

The fact that oil producing countries in the Middle East provide more than 50% of the world's consumption (El-Genk, 2008) is indicative for the low diversification of energy sources and the accompanying risks on smooth energy supply. The diversification that is offered by the alternative supplies from Russia and Africa cannot provide a sound solution for a supply disruption that may occur in the Middle East region. An overview of the oil market and the related risks and incidents clearly indicate that the risks associated to energy supply are many. War and civil conflicts might have been replaced, to some extent, by weather conditions and monopolistic practices, but they are still playing a crucial factor in energy supply (Doukas *et al.*, 2008). Therefore, the high dependency that most countries have in energy imports made essential for the policy makers to focus on the concept of security of energy supply. In this context, there is a need to assess the current energy system and the risks of energy disruptions in order to better design and adopt the required policies.

Uncertainty about future demand for oil, which will influence how quickly the remaining oil is used, contributes to the uncertainty about the timing of peak oil production. It is very likely that crude oil will continue to be a major source of energy well into the future and world consumption of petroleum products may even grow during the nest four decades.

Future world oil demand will depend on such uncertain factors as world economic growth, future government policy, and consumer choices.

Environmental concerns about gasoline's emissions of carbon dioxide, which is a greenhouse gas, may encourage future reductions in oil demand if these concerns are translated into policies that promote biofuels.

Consumer choices about conservation also can affect oil demand and thereby influence the timing of a peak. For example, if consumers in the United States were to purchase more fuel-efficient vehicles in greater numbers, this could reduce future oil demand in the United States, potentially delaying a time at which oil supply is unable to keep pace with oil demand. Such uncertainties that lead to changes in future oil demand ultimately make estimates of the timing of a peak uncertain. Specifically, using future annual increases in world oil consumption, ranging from 0%, to represent no increase, to 3%, to represent a large increase, and out of the various scenarios examined, it may be up to 75 years for when the peak may occur.

Factors that affect oil exploration and production also create uncertainty about the rate of production decline and the timing of the peak. The rate of decline after a peak is an important consideration because a decline that is more abrupt will likely have more adverse economic consequences than a decline that is less abrupt.

Consumer actions could help mitigate the consequences of a near-term peak and decline in oil production through demand-reducing behaviors such as carpooling; teleworking; and eco-driving measures, such as proper tire inflation and slower driving speeds. These energy savings come at some cost of convenience and productivity, and limited research has been done to estimate potential fuel savings associated with such efforts. However, estimates by the United States Department of Energy indicate that teleworking could reduce total fuel consumption in the by 1 to 4%, depending on whether teleworking is undertaken for two days per week or the full five-day week, respectively.

If the peak occurs in the more distant future or the decline following a peak is less severe, alternative technologies have a greater potential to mitigate the consequences. The United States Department of Energy projects that the alternative technologies have the potential to displace up to the equivalent of 34% of annual U.S. consumption of petroleum products in the 2025 through 2030 time frame. However, the United States Department of Energy also considers these projections optimistic because the assumption is that sufficient time and effort are dedicated to the development of these technologies to overcome the challenges they face.

More specifically, the United States Department of Energy assumes sustained high oil prices above $50 per barrel as a driving force. The level of effort dedicated to overcoming challenges to alternative technologies will depend in part on the price of oil, with higher oil prices creating incentives to develop alternatives. High oil prices also enhance consumer interest in alternatives that consume less oil. For example, new purchases of light trucks, SUVs, and minivans usually decline during a period of increasing gasoline prices.

The prospect of a peak in oil production presents problems of global proportion whose consequences will depend critically on our preparedness. The consequences would be most dire if a peak occurred soon, without warning, and were followed by a sharp decline in oil production because alternative energy sources, particularly for transportation, are not yet available in large quantities. Such a peak would require sharp reductions in oil consumption, and the competition for increasingly scarce energy would drive up prices, possibly to unprecedented levels, causing severe economic damage. While these consequences would be felt globally, the United States, being the largest consumer of oil and one of the nation's most heavily dependent on oil for transportation, is especially vulnerable.

The subject of energy security has been for many years an important concern among energy policy makers. The devastating short and long term effects of the oil crisis of 1973 in the global economy made clear since then that the need to guarantee the availability of energy resource supply in a sustainable and timely manner with the energy price being at a level that will not adversely affect the economic performance the European continent is of utmost importance (Asia Pacific Research Centre 2007). The continuous destabilization of the Middle East, growing fears about further military intervention in this fragile geopolitical area, environmental catastrophes, the advent of organized terrorist operations across the globe, political risks and legal reforms have profoundly increased the possibility of potential energy disruptions that will have detrimental effects, considering the dependence of Europe to external energy suppliers.

The popularity of the *energy* risk-premium concept has led to the formulation of a vast pool of knowledge encompassing an abundance of derivatives models. Traders have the ability to hedge against various risks and create risk neutral portfolios using a diversified mix of energy derivative securities. However, up to now the risk-premium concept has not been used in the energy domain to quantify an energy security indicator. The main reason for this is that current techniques used in the energy domain do not incorporate the necessary probabilistic models that reflect on risk parameters associated with rare catastrophic events that cause adverse movements on the sport price of the underlying instrument.

A catastrophic event in general is an event which has severe losses, injury, or property damage, affects large population of exposures and is caused by natural or handmade events (Fisher and Kist 2001). Examples of catastrophic events include natural disasters, such as hurricanes, earthquakes, floods, and terrorist attacks. Over the last 20 years, natural catastrophes have been happening with increasing intensity.

Catastrophic events in the energy context have a slightly different meaning. They can be events with low frequency of occurrence that cause the spot price of the energy commodity to soar. Usually a price increase due to the catastrophic event does not have a lasting effect and the spot price tends to return to or close to its initial value. To combat such events, it will be necessary for the non-oil producing nations to commence development of sources of energy other than oil.

Once all risk indicators associated with catastrophic events have been identified and properly estimated in terms of frequency of occurrence and impact in the underlying spot price, then the respective premium can be calculated under the common assumptions of derivatives pricing.

In the longer term, there are many possible alternatives to using oil, including using biofuels and improving automotive fuel efficiency, but these alternatives will require large investments, and in some cases, major changes in infrastructure or break-through technological advances. In the past, the private sector has responded to higher oil prices by investing in alternatives, and it is doing so now. However, investment, however, is determined largely by price expectations, so unless high oil prices are sustained, we cannot expect private investment in alternatives to continue at current levels. If a peak were anticipated, oil prices would rise, signaling industry to increase efforts to develop alternatives and consumers of energy to conserve and look for more energy-efficient products.

Finally, with the onset of the 21st century, petroleum technology is driven by the increasing supply of heavy oils with decreasing quality and the fast increases in the demand for clean and ultra-clean vehicle fuels and petrochemical raw materials. As feedstocks to refineries change, there must be an accompanying change in refinery technology. This means a movement from conventional means of refining heavy feedstocks, typically using coking technologies to more innovative processes, including hydrogen management, that will produce the ultimate amounts liquid fuels from the feedstock and maintain emissions within environmental compliance (Penning, 2001; Lerner, 2002; Davis and Patel, 2004; Speight, 2008; Speight, 2011).

To meet the challenges from changing restructured over the years from simple crude trends in product slate and the stringent distillation operations into increasingly specifications imposed by environmental complex chemical operations involving legislation, the refining industry in the near transformation of crude oil into a variety of future will become increasingly flexible and refined products with specifications that meet innovative with new processing schemes, users requirements.

During the next 20 to 30 years, the evolution future of petroleum refining and the current refinery layout will be primarily on process modification with some new innovations coming on-stream. The industry will move predictably on to deep conversion of heavy feedstocks, higher hydrocracking and hydrotreating capacity, and more efficient processes.

High conversion refineries will move to gasification of feedstocks for the development of alternative fuels and to enhance equipment usage. A major trend in the refining industry market demand for refined products will be in synthesizing fuels from simple basic reactants (e.g. synthesis gas) when it becomes uneconomical to produce super clean transportation fuels through conventional refining processes. Fischer-Tropsch plants together with IGCC systems will be integrated with, or even into refineries, which will offer the advantage of high quality products.

In summary, the petroleum industry is indeed at the verge of a major decision period with the onset of processing high volumes of heavy crude oil and residua. Several technology breakthroughs have made this possible, but many technical challenges remain and some are being met, including the production of fuels derived from sources other than petroleum (Speight, 2008; Høygaard Michaelsen *et al.*, 2009; Luce, 2009; Spewight, 2011).

9.7 References

Anderson, L.L., and Tillman, D.A. 1979. Synthetic Fuels from Coal: Overview and Assessment. John Wiley and Sons Inc., New York.

Andrews, A. 2006. Oil Shale: History, Incentives, and Policy. Specialist, Industrial Engineering and Infrastructure Policy Resources, Science, and Industry Division. Congressional Research Service, the Library of Congress, Washington, DC.

Annual Energy Outlook. 2008. Annual Energy Outlook 2008 with Projections to 2030. Report No. DOE/EIA-0383(2008). Energy Information Administration, Department of Energy, Washington, DC. June.

Argonne. 1990. Environmental Consequences of, and Control Processes for, Energy Technologies. Argonne National Laboratory. Pollution Technology Review No. 181. Noyes Data Corp., Park Ridge, New Jersey. Chapter 6.

ASTM. 2008. Annual Book of Standards. American Society for Testing and Materials, West Conshohocken, Pennsylvania.

Attanasi, E.D. 1999. Economics of Undiscovered Oil in the 1002 Area of the Arctic National Wildlife Refuge. In The Oil and gas resource Potential 1002 Area of the Arctic National Wildlife Refuge. Open File Report 98–34. United States Geological Survey, Reston Virginia.

Bala, B. K. 2005. Studies on Biodiesel from the Transformation of Vegetable Oils. Energy Education Sci. Technol. 15: 1–45.

Baldwin, R.M. 2002. Oil Shale: A Brief Technical Overview. Colorado School of Mines, Golden, Colorado. July.

Bartis, J.T., LaTourrette, T., and Dixon, L. 2005. Oil Shale Development in the United States: Prospects and Policy Issues. Prepared for the National Energy Technology of the United States Department of Energy. Rand Corporation, Santa Monica, California.

Batchelder, H.R. 1962. In Advances in Petroleum Chemistry and Refining. J.J. McKetta Jr. (Editor). Interscience Publishers Inc., New York. Volume V. Chapter 1.

Bockey, D. 2006. Potentials for Raw Materials for the Production of Biodiesel: An Analysis. Union zur Förderung von Oel- und Proteinpflanzen e. V., Berlin Claire-Waldoff-Strasse 7, 10117 Berlin.

Bourne, J.K. 2007. Biofuels: Boon or Boondoggle. National Geographic Magazine. 212(4): 38–59.

Brown, R.C. 2003. Biorenewable Resources: Engineering New Products from Agriculture. Iowa State Press, Ames, Iowa.

Berthelot, M. 1869. Bull. Soc. Chim. France. 11: 278.

BP. 2010. Statistical Review of World Energy 2007. British Petroleum Company Ltd., London, England. June.

Campbell, C.J., and Laherrère, J.H. 1998. The End of Cheap Oil. Scientific American, March.

Crocker, M., and Crofcheck, C. 2006. Reducing national dependence on imported oil. Energeia Vol. 17, No. 6. Center for Applied Energy Research, University of Kentucky, Lexington, Kentucky.

Davis, R.A., and Patel, N.M. 2004. Refinery Hydrogen Management. Petorleum Technology Quarterly, Spring: 29–35.

Davis, G.W. 2006. Using E85 in Vehicles. In Alcoholic Fuels. S. Minteer (Editor). CRC-Taylor & Francis, Boca Raton, Florida. Chapter 8.

Demirbaş, A. 2005. Bioethanol from Cellulosic Materials: A Renewable Motor Fuel from Biomass. Energy Sources. 27: 327–337.

DOE 2004a. Strategic Significance of America's Oil Shale Reserves, I. Assessment of Strategic Issues, March. http://www.fe.doe.gov/programs/reserves/publications

DOE 2004b. Strategic Significance of America's Oil Shale Reserves, II. Oil Shale Resources, Technology, and Economics; March. http://www.fe.doe.gov/programs/reserves/publications

DOE 2004c. America's Oil Shale: A Roadmap for Federal Decision Making; United States Department of Energy, Office of US Naval Petroleum and Oil Shale Reserves. http://www.fe.doe.gov/programs/reserves/publications

Freeman, S.D. 2007. Winning Our Energy Independence: An Insider Show How. Gibbs Smith, Salt Lake City, Utah.

Gordon, 2007. Crude Oil: Uncertainty about Future Oil Supply Makes It Important to Develop a Strategy for Addressing a Peak and Decline in Oil Production. Report No. GAO-07-283. Report to Congressional Requesters. General Accountability Office, Washington, DC. February.

Gorin, E. 1981. In Chemistry of Coal Utilization. Second Supplementary Volume. M.A. Elliott (Editor). John Wiley & Sons Inc., New York. Chapter 27.

Høygaard Michaelsen, N., Egeberg, R., and Nyström, S. 2009. Consider New Technology to Produce Renewable Diesel. Hydrocarbon Processing, 88(2): 41–44.

Karanikas; J.M., de Rouffignac, E.P., Vinegar; H.J. (Houston, TX), and Wellington, S. 2005. In Situ Thermal Processing of An Oil Shale Formation While Inhibiting Coking. United States Patent 6,877,555, April 12.

Knothe, G., Krahl, J., and Van Gerpen, J. 2005. The Biodiesel Handbook. AOCS Press, Champaign, Illinois.

Lee, S., Speight, J.G., and Loyalka, S. 2007. Handbook of Alternative Fuel Technologies. CRC Press, Taylor & Francis Group, Boca Raton, Florida. 2007.

Lerner, B. 2002. The Future of Refining. Hydrocarbon Engineering. September.

Luce, G.W. 2009. Renewable Energy Solutions for and Enegy hungry World. Hydrocarbon Processing, 88(2): 19–22.

Lucia, L.A., Argyropoulos, D.S., Adamopoulos, L. and Gaspar, A.R. 2006. Chemicals and energy from biomass. *Can. J. Chem.* 84:960–970.

McNeil Technologies Inc. 2005. Colorado Agriculture IOF: Technology Assessments Liquid Fuels. Prepared Under State of Colorado Purchase Order # 01-336. Governor's Office of Energy Conservation and Management, Denver, Colorado.

Minteer, S. 2006. Alcoholic Fuels: An Overview. In Alcoholic Fuels. S. Minteer (Editor). CRC-Taylor & Francis, Boca Raton, Florida. Chapter 1.

Narayan, R. 2007. Rationale, Drivers, Standards, and Technology for Biobased Materials. In Renewable Resources and Renewable energy: A Global Challenge. M. Graziana and P. Fornaserio (Editors). CRC Press-Taylor & Francis Group, Boca Raton, Florida. Chapter 1.

Nersesian, R.L. 2007. Energy for the 21st Century: A Comprehensive Guide to Conventional and Alternative Fuel Sources. M.S. Sharpe Publishers, Armonk, New York.

Penning, R.T. 2001. Petroleum Refining: A Look at The Future. Hydrocarbon Processing, 80(2): 45–46.

Pimentel, D., and Pimentel, M. 2006. Global Environmental Resources Versus World Population Growth. Ecological Economics. 59: 195–198.

Scouten, C. 1990. In Fuel Science and Technology Handbook. J.G. Speight (Editor). Marcel Dekker Inc., New York.

Smith, I.M. 2006. Management of FGD Residues. Center for Applied Energy Research, University of Kentucky, Lexington, Kentucky.

Speight, J.G. 1990. In Fuel Science and Technology Handbook. J.G. Speight (Editor). Marcel Dekker Inc., New York. Chapter 33.

Speight, J.G. 1994. The Chemistry and Technology of Coal. 2nd Edition. Marcel Dekker inc., New York.

Speight, J.G. 2007. The Chemistry and Technology of Petroleum. 4th Edition. CRC-Taylor and Francis Group, Boca Raton, Florida.

Speight, J.G. 2008. Handbook of Synthetic Fuels Handbook: Properties, Processes, and Performance. McGraw-Hill, New York.

Speight, J.G. 2009. Enhanced Recovery Methods for Heavy Oil and Tar Sands, Gulf Publishing Company, Houston, Texas.

Speight, J.G. 2011. The Refinery of the Future, Gulf Professional Publishing, Elsevier, Oxford, United Kingdom.

Steinhubl, A., Wilczynski, H., Pettit, J., and Click, C. 2009. Unconventional resources to Keep Pivotal Supply Role. Oil & Gas Journal. 107(4): 18–20.

Stranges, A.N. 1983. J. Chem. Educ. 60: 617.

Stranges, A.N. 1987. Fuel Processing Technology. 16: 205.

Tsai, W.T., Lee, M.K., and Chang, Y.M. 2007. Fast Pyrolysis of Rice h\Husk: Product Yields and Compositions. Bioresource Technology 98:22–28.

Whitehurst, D.D., Mitchell, T.O., and Farcasiu, M. 1980. Coal Liquefaction: The Chemistry and Technology of Thermal Processes. Academic Press Inc., New York.

Worldwatch Institute. 2006. Biofuels for Transport: Global Potential and Implications for Energy and Agriculture. Prepared by the Worldwatch Institute for the German ministry of Food, Agriculture, and Consumer Protection (BMELV) in Coordination with the German Agency for

Technical Cooperation (GTZ) and the German Agency of Renewable Resources (FNR). Earthscan, London, UK.

Wright, L., Boundy, R, Perlack, R., Davis, S., and Saulsbury. B. 2006. Biomass Energy Data Book: Edition 1. Office of Planning, Budget and Analysis, Energy Efficiency and Renewable Energy, United States Department of Energy. Contract No. DE-AC05-00OR22725. Oak Ridge National Laboratory, Oak Ridge, Tennessee.

Conversion Factors

1 acre = 43,560 sq ft
1 acre foot = 7758.0 barrels
1 atmosphere = 760 mm Hg = 14.696 psia = 29.91 in. Hg
1 atmosphere = 1.0133 bars = 33.899 ft. H_2O
1 barrel (oil) = 42 gal = 5.6146 *cu* ft
1 barrel (water) = 350 lb at 60°F
1 barrel per day = 1.84 cu cm per second
1 Btu = 778.26 ft-lb
1 centipoise × 2.42 = lb mass/(ft) (hr), viscosity
1 centipoise × 0.000672 = lb mass/(ft) (sec), viscosity
1 cubic foot = 28,317 cu cm = 7.4805 gal
Density of water at 60°F = 0.999 gram/cu cm = 62.367 lb/cu ft = 8.337 lb/gal
1 gallon = 231 cu in. = 3,785.4 cu cm = 0.13368 cu ft
1 horsepower-hour – 0.7457 kwhr = 2544.5 Btu
1 horsepower = 550 ft-lb/sec = 745.7 watts
1 inch = 2.54 cm
1 meter = 100 cm = 1,000 mm = 10 microns = 10 angstroms (A)
1 ounce = 28.35 grams
1 pound = 453.59 grams = 7,000 grains
1 square mile = 640 acres

Glossary

3-D seismic: Seismic data that is acquired and processed to yield a three-dimensional picture of the subsurface.

A

Abandonment costs: Costs associated with abandoning a well or production facility. Such costs are specified in the authority for expenditure and typically cover the plugging of wells; removal of well equipment, production tanks and associated installations; and surface remediation.

Absorber: See Absorption tower.

Absorption gasoline: Gasoline extracted from natural gas or refinery gas by contacting the absorbed gas with oil and subsequently distilling the gasoline from the higher-boiling components.

Absorption oil: Oil used to separate the heavier components from a vapor mixture by absorption of the heavier components during intimate contacting of the oil and vapor; used to recover natural gasoline from wet gas.

Absorption plant: A plant for recovering the condensable portion of natural or refinery gas, by absorbing the higher boiling hydrocarbons in an absorption oil, followed by separation and fractionation of the absorbed material.

Absorption tower: A tower or column, which promotes contact between a rising gas and a falling liquid so that part of the gas may be dissolved in the liquid.

Acid catalyst: A catalyst having acidic character; the aluminas are examples of such catalysts.

Acid deposition: Acid rain; a form of pollution depletion in which pollutants, such as nitrogen oxides and sulfur oxides, are transferred from the atmosphere to soil or water; often referred to as atmospheric self-cleaning. The pollutants usually arise from the use of fossil fuels.

Acid rain: The precipitation phenomenon that incorporates anthropogenic acids and other acidic chemicals from the atmosphere to the land and water (see Acid deposition).

Acoustic log: A display of travel time of acoustic waves versus depth in a well; the term is commonly used as a synonym for a sonic log; some acoustic logs display velocity.

Afterflow: The flow associated with wellbore storage following a surface shut-in. When a well is first shut in at the surface, flow from the formation into the bottom of the wellbore continues unabated until compression of the fluids in the wellbore causes the downhole pressure to rise; if the wellbore fluid is highly compressible and the well rate is low, the afterflow period can be long; conversely, high-rate wells producing little gas have negligible afterflow periods.

Air-blown asphalt: Asphalt produced by blowing air through residua at elevated temperatures.

Air pollution: The discharge of toxic gases and particulate matter introduced into the atmosphere, principally as a result of human activity.

Alcohol: The family name of a group of organic chemical compounds composed of carbon, hydrogen, and oxygen. The molecules in the series vary in chain length and are composed of a hydrocarbon plus a hydroxyl group. Alcohol includes methanol and ethanol.

Alkylation: A process for manufacturing high octane blending components used in unleaded petrol or gasoline.

Alumina (Al_2O_3): Used in separation methods as an adsorbent and in refining as a catalyst.

American Society for Testing and Materials (ASTM): The official organization in the United States for designing standard tests for petroleum and other industrial products.

Anaerobic digestion: Decomposition of biological wastes by micro-organisms, usually under wet conditions, in the absence of air (oxygen), to produce a gas comprising mostly methane and carbon dioxide.

Annual removals: The net volume of growing stock trees removed from the inventory during a specified year by harvesting, cultural operations such as timber stand improvement, or land clearing.

API gravity: A measure of the lightness or heaviness of petroleum that is related to density and specific gravity: API = (141.5/sp gr @ 60°F) - 131.5.

Aquifer: A subsurface rock interval that will produce water; often the underlay of a petroleum reservoir.

Aromatics: A range of hydrocarbons which have a distinctive sweet smell and include benzene and toluene' occur naturally in petroleum and are also extracted as a petrochemical feedstock, as well as for use as solvents.

Asphalt: The nonvolatile product obtained by distillation and treatment of an asphaltic crude oil; a manufactured product.

Asphalt cement: Asphalt especially prepared as to quality and consistency for direct use in the manufacture of bituminous pavements.

Asphaltene (asphaltene constituents): The brown to black powdery material produced by treatment of petroleum, heavy oil, bitumen, or residuum with a low-boiling liquid hydrocarbon.

B

Bank: The concentration of oil (oil bank) in a reservoir that moves cohesively through the reservoir.

Barrel (bbl): The unit of measure used by the petroleum industry; equivalent to approximately forty-two US gallons or approximately thirty four (33.6) Imperial gallons or 159 liters; 7.2 barrels are equivalent to one tonne of oil (metric).

Barrel of oil equivalent (boe): The amount of energy contained in a barrel of crude oil, i.e. approximately 6.1 GJ (5.8 million Btu), equivalent to 1,700 kWh.

Bbl: A barrel of 42 U.S. gallons of oil.

Bcf: Billion cubic feet of natural gas.

Bcfe: Billion cubic feet equivalent, determined using the ratio of six Mcf of natural gas to one Bbl of crude oil, condensate or natural gas liquids.

Benzene: A colorless aromatic liquid hydrocarbon (C_6H_6).

Benzin: A refined light naphtha used for extraction purposes.

Benzine: An obsolete term for light petroleum distillates covering the gasoline and naphtha range; see Ligroine.

Benzol: The general term which refers to commercial or technical (not necessarily pure) benzene; also the term used for aromatic naphtha.

Billion: 1×10^9.

Biochemical conversion: The use of fermentation or anaerobic digestion to produce fuels and chemicals from organic sources.

Biodiesel: A fuel derived from biological sources that can be used in diesel engines instead of petroleum-derived diesel; through the process of transesterification, the triglycerides in the biologically derived oils are separated from the glycerin, creating a clean-burning, renewable fuel.

Bioenergy: Useful, renewable energy produced from organic matter - the conversion of the complex carbohydrates in organic matter to energy; organic matter may either be used directly as a fuel, processed into liquids and gasses, or be a residual of processing and conversion.

Bioethanol: Ethanol produced from biomass feedstocks; includes ethanol produced from the fermentation of crops, such as corn, as well as cellulosic ethanol produced from woody plants or grasses.

Biofuels: A generic name for liquid or gaseous fuels that are not derived from petroleum based fossils fuels or contain a proportion of non fossil fuel; fuels produced from plants, crops such as sugar beet, rape seed oil or re-processed vegetable oils or fuels made from gasified biomass; fuels made from renewable biological sources and include ethanol, methanol, and biodiesel; sources include, but are not limited to: corn, soybeans, flaxseed, rapeseed, sugarcane, palm oil, raw sewage, food scraps, animal parts, and rice.

Biogas: A combustible gas derived from decomposing biological waste under anaerobic conditions. Biogas normally consists of 50 to 60 percent methane. See also landfill gas.

Biomass: Any organic matter that is available on a renewable or recurring basis, including agricultural crops and trees, wood and wood residues, plants (including aquatic plants), grasses, animal manure, municipal residues, and other residue materials. Biomass is generally produced in a sustainable manner from water and carbon dioxide by photosynthesis. There are three main categories of biomass - primary, secondary, and tertiary.

Biomass to liquid (BTL): The process of converting biomass to liquid fuels. Hmm, that seems painfully obvious when you write it out.

Biopower: The use of biomass feedstock to produce electric power or heat through direct combustion of the feedstock, through gasification and then combustion of the resultant gas, or through other thermal conversion processes. Power is generated with engines, turbines, fuel cells, or other equipment.

Biorefinery: A facility that processes and converts biomass into value-added products. These products can range from biomaterials to fuels such as ethanol or important feedstocks for the production of chemicals and other materials.

Bitumen: On occasion, referred to as native asphalt, and extra heavy oil; a naturally occurring material that has little or no mobility under reservoir conditions and which cannot be recovered through a well by conventional oil well production methods including currently used enhanced recovery techniques; current methods involve mining for bitumen recovery.

Bituminous: Containing bitumen or constituting the source of bitumen.

Bituminous rock: See Bituminous sand.

Bituminous sand: A formation in which the bituminous material (see Bitumen) is found as a filling in veins and fissures in fractured rock or impregnating relatively shallow sand, sandstone, and limestone strata; a sandstone reservoir that is impregnated with a heavy, viscous black petroleum-like material that cannot be retrieved through a well by conventional production techniques.

Black liquor: Solution of lignin-residue and the pulping chemicals used to extract lignin during the manufacture of paper.

BOE: Barrel of oil equivalent.

Boiling point: A characteristic physical property of a liquid at which the vapor pressure is equal to that of the atmosphere and the liquid is converted to a gas.

Boiling range: The range of temperature, usually determined at atmospheric pressure in standard laboratory apparatus, over which the distillation of an oil commences, proceeds, and finishes.

Bone dry: Having zero percent moisture content. Wood heated in an oven at a constant temperature of 100°C (212°F) or above until its weight stabilizes is considered bone dry or oven dry.

Bottomhole shut-in: A well shut in slightly above the producing formation by use of special downhole tools containing a valve that can be preprogrammed or controlled from the surface; this practice is commonly associated with drillstem tests; technology exists to employ bottomhole shut-in in suitably equipped completed wells.

Bottoming cycle: A cogeneration system in which steam is used first for process heat and then for electric power production.

British thermal unit - (Btu): A non-metric unit of heat, still widely used by engineers; One Btu is the heat energy needed to raise the temperature of one pound of water from 60°F to 61°F at one atmosphere pressure. 1 Btu = 1055 joules (1.055 kJ).

Bunker: A storage tank.

Butanol: Though generally produced from fossil fuels, this four-carbon alcohol can also be produced through bacterial fermentation of alcohol.

C

C_1, C_2, C_3, C_4, C_5 fractions: a common way of representing fractions containing a preponderance of hydrocarbons having 1, 2, 3, 4, or 5 carbon atoms, respectively, and without reference to hydrocarbon type.

Carbon dioxide (CO_2): A product of combustion that acts as a greenhouse gas in the Earth's atmosphere, trapping heat and contributing to climate change.

Carbonization: The conversion of an organic compound into char or coke by heat in the substantial absence of air; often used in reference to the destructive distillation (with simultaneous removal of distillate) of coal.

Carbon monoxide (CO): A lethal gas produced by incomplete combustion of carbon-containing fuels in internal combustion engines. It is colorless, odorless, and tasteless. (As in flavorless, we mean, though it's also been known to tell a bad joke or two.)

Carbon sink: A geographical area whose vegetation and/or soil soaks up significant carbon dioxide from the atmosphere. Such areas, typically in tropical regions, are increasingly being sacrificed for energy crop production.

Catalyst: A substance that accelerates a chemical reaction without itself being affected or consumed. In refining, catalysts are used in the cracking process to produce blending components for fuels.

Catalyst selectivity: The relative activity of a catalyst with respect to a particular compound in a mixture, or the relative rate in competing reactions of a single reactant.

Cetane number: A measure of the ignition quality of diesel fuel; the higher the number the more easily the fuel is ignited under compression.

CFR: Code of Federal Regulations; Title 40 (40 CFR) contains the regulations for protection of the environment.

Closed-loop biomass: Crops grown, in a sustainable manner, for the purpose of optimizing their value for bioenergy and bioproduct uses. This includes annual crops such as maize and wheat, and perennial crops such as trees, shrubs, and grasses such as switch grass.

Cloud point: The temperature at which paraffin wax or other solid substances begin to crystallise or separate from the solution, imparting a cloudy appearance to the oil when the oil is chilled under prescribed conditions.

Coal tar: The specific name for the tar (*q.v.*) produced from coal.

Coal tar pitch: The specific name for the pitch (*q.v.*) produced from coal.

Coarse materials: Wood residues suitable for chipping, such as slabs, edgings, and trimmings.

Coking: A thermal method used in refineries for the conversion of bitumen and residua to volatile products and coke (see Delayed coking and Fluid coking).

Completion: The installation of permanent equipment for the production of oil or gas.

Conformance: the uniformity with which a volume of the reservoir is swept by injection fluids in area and vertical directions.

Conventional crude oil (conventional petroleum): Crude oil that is pumped from the ground and recovered using the energy inherent in the reservoir; also recoverable by application of secondary recovery techniques.

Cord: A stack of wood comprising 128 cubic feet (3.62 m^3); standard dimensions are $4 \times 4 \times 8$ feet, including air space and bark. One cord contains approx. 1.2 U.S. tons (oven-dry) = 2400 pounds = 1089 kg.

Cracking: A secondary refining process that uses heat and/or a catalyst to break down high molecular weight chemical components into lower molecular weight products which can be used as blending components for fuels.

Craig-Geffen-Morse method: A method for predicting oil recovery by water flood.

Cropland: Total cropland includes five components: cropland harvested, crop failure, cultivated summer fallow, cropland used only for pasture, and idle cropland.

Cropland pasture: Land used for long-term crop rotation. However, some cropland pasture is marginal for crop uses and may remain in pasture indefinitely. This category also includes land that was used for pasture before crops reached maturity and some land used for pasture that could have been cropped without additional improvement.

Crude oil: See Petroleum.

Cull tree: A live tree, 5.0 inches in diameter at breast height (dbh) or larger that is non-merchantable for saw logs now or prospectively because of rot, roughness, or species. (See definitions for rotten and rough trees.)

Cultivated summer fallow: Cropland cultivated for one or more seasons to control weeds and accumulate moisture before small grains are planted.

D

Delayed coking: A coking process in which the thermal reactions are allowed to proceed to completion to produce gaseous, liquid, and solid (coke) products.

Density: The mass (or weight) of a unit volume of any substance at a specified temperature; see also Specific gravity.

Desulfurization: The removal of sulfur or sulfur compounds from a feedstock.

Development well: A well drilled within the proved area of an oil or gas reservoir to the depth of a stratigraphic horizon known to be productive.

Diesel engine: Named for the German engineer Rudolph Diesel, this internal-combustion, compression-ignition engine works by heating fuels and causing them to ignite; can use either petroleum or bio-derived fuel.

Diesel fuel: A distillate of fuel oil that has been historically derived from petroleum for use in internal combustion engines; also derived from plant and animal sources.

Diesel, Rudolph: German inventor famed for fashioning the diesel engine, which made its debut at the 1900 World's Fair; initially engine to run on vegetable-derived fuels.

Digester: An airtight vessel or enclosure in which bacteria decomposes biomass in water to produce biogas.

Direct-injection engine: A diesel engine in which fuel is injected directly into the cylinder.

Distillate: Any petroleum product produced by boiling crude oil and collecting the vapors produced as a condensate in a separate vessel, for example gasoline (light distillate), gas oil (middle distillate), or fuel oil (heavy distillate).

Distillation: The primary distillation process which uses high temperature to separate crude oil into vapor and fluids which can then be fed into a distillation or fractionating tower.

Downdraft gasifier: A gasifier in which the product gases pass through a combustion zone at the bottom of the gasifier.

Dutch oven furnace: One of the earliest types of furnaces, having a large, rectangular box lined with firebrick (refractory) on the sides and top; commonly used for burning wood.

E

E85: An alcohol fuel mixture containing 85 percent ethanol and 15 percent gasoline by volume, and the current alternative fuel of choice of the U.S. government.

Effluent: The liquid or gas discharged from a process or chemical reactor, usually containing residues from that process.

Emissions: Substances discharged into the air during combustion, e.g., all that stuff that comes out of your car.

Emissions: Waste substances released into the air or water.

Energy crops: Crops grown specifically for their fuel value; include food crops such as corn and sugarcane, and nonfood crops such as poplar trees and switch grass.

Energy balance: The difference between the energy produced by a fuel and the energy required to obtain it through agricultural processes, drilling, refining, and transportation.

Energy crops: Agricultural crops grown specifically for their energy value.

Energy-efficiency ratio: A number representing the energy stored in a fuel as compared to the energy required to produce, process, transport, and distribute that fuel.

Enhanced recovery: Methods that usually involves the application of thermal energy (e.g., steam flooding) to oil recovery from the reservoir.

Enhanced oil recovery (EOR) process: A method for recovering additional oil from a petroleum reservoir beyond that economically recoverable by conventional primary and secondary recovery methods. EOR methods are usually divided into three main categories: (1) chemical flooding: injection of water with added chemicals into a petroleum reservoir. The chemical processes include: surfactant flooding, polymer flooding, and alkaline flooding, (2) miscible flooding: injection into a petroleum reservoir of a material that is miscible, or can become miscible, with the oil in the reservoir. Carbon dioxide, hydrocarbons, and nitrogen are used, (3) thermal recovery: injection of steam into a petroleum reservoir, or propagation of a combustion zone through a reservoir by air or oxygen- enriched air injection. The thermal processes include: steam drive, cyclic steam injection, and in situ combustion.

Ethanol (ethyl alcohol, alcohol, or grain-spirit): A clear, colorless, flammable oxygenated hydrocarbon; used as a vehicle fuel by itself (E100 is 100% ethanol by volume), blended with gasoline (E85 is 85% ethanol by volume), or as a gasoline octane enhancer and oxygenate (10% by volume).

Exploratory well: A well drilled to find and produce oil or gas in an unproved area, to find a new reservoir in a field previously found to be productive of oil or gas in another reservoir, or to extend a known reservoir.

F

FCC: Fluid catalytic cracking.

FCCU: Fluid catalytic cracking unit.

Feedstock: The biomass used in the creation of a particular biofuel (e.g., corn or sugarcane for ethanol, soybeans or rapeseed for biodiesel).

Fermentation: Conversion of carbon-containing compounds by micro-organisms for production of fuels and chemicals such as alcohols, acids or energy-rich gases.

Fiber products: Products derived from fibers of herbaceous and woody plant materials; examples include pulp, composition board products, and wood chips for export.

Fine materials: Wood residues not suitable for chipping, such as planer shavings and sawdust.

Flexible-fuel vehicle (flex-fuel vehicle): A vehicle that can run alternately on two or more sources of fuel; includes cars capable of running on gasoline and gasoline/ethanol mixtures, as well as cars that can run on both gasoline and natural gas.

Fluid coking: A continuous fluidised solids process that cracks feed thermally over heated coke particles in a reactor vessel to gas, liquid products, and coke.

Fluidized-bed boiler: A large, refractory-lined vessel with an air distribution member or plate in the bottom, a hot gas outlet in or near the top, and some provisions for introducing fuel; the fluidized bed is formed by blowing air up through a layer of inert particles (such as sand or limestone) at a rate that causes the particles to go into suspension and continuous motion.

Fly ash: Small ash particles carried in suspension in combustion products.

Forest land: Land at least 10% stocked by forest trees of any size, including land that formerly had such tree cover and that will be naturally or artificially regenerated; includes transition zones, such as areas between heavily forested and non-forested lands that are at least 10% stocked with forest trees and forest areas adjacent to urban and built-up lands; also included are pinyon-juniper and chaparral areas; minimum area for classification of forest land is 1 acre.

Forest residues: Material not harvested or removed from logging sites in commercial hardwood and softwood stands as well as material resulting from forest management operations such as pre-commercial thinnings and removal of dead and dying trees.

Forest health: A condition of ecosystem sustainability and attainment of management objectives for a given forest area; usually considered to include green trees, snags, resilient stands growing at a moderate rate, and endemic levels of insects and disease.

Formation: An interval of rock with distinguishable geologic characteristics.

Fossil fuel: Solid, liquid, or gaseous fuels formed in the ground after millions of years by chemical and physical changes in plant and animal residues under high temperature and pressure. Oil, natural gas, and coal are fossil fuels.

Fossil fuel resources: a gaseous, liquid, or solid fuel material formed in the ground by chemical and physical changes (diagenesis, q.v.) in plant and animal residues over geological time; natural gas, petroleum, coal, and oil shale.

Fuel cell: A device that converts the energy of a fuel directly to electricity and heat, without combustion.

Fuel cycle: The series of steps required to produce electricity. The fuel cycle includes mining or otherwise acquiring the raw fuel source, processing and cleaning the fuel, transport, electricity generation, waste management and plant decommissioning.

Fuel oil: A heavy residue, black in color, used to generate power or heat by burning in furnaces.

Fuel treatment evaluator (FTE): A strategic assessment tool capable of aiding the identification, evaluation, and prioritization of fuel treatment opportunities.

Fuel wood: Wood used for conversion to some form of energy, primarily for residential use.

Furnace: An enclosed chamber or container used to burn biomass in a controlled manner to produce heat for space or process heating.

G

Gasification: A chemical or heat process used to convert carbonaceous material (such as coal, petroleum, and biomass) into gaseous components such as carbon monoxide and hydrogen.

Gasifier: A device for converting solid fuel into gaseous fuel; in biomass systems, the process is referred to as pyrolitic distillation.

Gasohol: A mixture of 10% anhydrous ethanol and 90% gasoline by volume; 7.5% anhydrous ethanol and 92.5% gasoline by volume; or 5.5% anhydrous ethanol and 94.5% gasoline by volume.

Gas to liquids (GTL): The process of refining natural gas and other hydrocarbons into longer-chain hydrocarbons, which can be used to convert gaseous waste products into fuels.

Gel point: The point at which a liquid fuel cools to the consistency of petroleum jelly.

Genetically modified organism (GMO): An organism whose genetic material has been modified through recombinant DNA technology, altering the phenotype of the organism to meet desired specifications.

Grassland pasture and range: All open land used primarily for pasture and grazing, including shrub and brush land types of pasture; grazing land with sagebrush and scattered mesquite; and all tame and native grasses, legumes, and other forage used for pasture or grazing; because of the diversity in vegetative composition, grassland pasture and range are not always clearly distinguishable from other types of pasture and range; at one extreme, permanent grassland may merge with cropland pasture, or grassland may often be found in transitional areas with forested grazing land.

Grease car: A diesel-powered automobile rigged post-production to run on used vegetable oil.

Greenhouse effect: The effect of certain gases in the Earth's atmosphere in trapping heat from the sun.

Greenhouse gases: Gases that trap the heat of the sun in the Earth's atmosphere, producing the greenhouse effect. The two major greenhouse gases are water vapor and carbon dioxide. Other greenhouse gases include methane, ozone, chlorofluorocarbons, and nitrous oxide.

Grid: An electric utility company's system for distributing power.

Gross: When used with respect to acres or wells, refers to the total acres or wells in which a company, individual, trust, or foundation has a working interest.

Growing stock: A classification of timber inventory that includes live trees of commercial species meeting specified standards of quality or vigor; cull trees are excluded.

H

Habitat: The area where a plant or animal lives and grows under natural conditions. Habitat includes living and non-living attributes and provides all requirements for food and shelter.

Hardwoods: Usually broad-leaved and deciduous trees.

Heating value: The maximum amount of energy that is available from burning a substance.

Heavy oil (heavy crude oil): Oil that is more viscous that conventional crude oil, has a lower mobility in the reservoir but can be recovered through a well from the reservoir by the application of a secondary or enhanced recovery methods.

Hectare: Common metric unit of area, equal to 2.47 acres. 100 hectares = 1 square kilometer.

Herbaceous: Non-woody type of vegetation, usually lacking permanent strong stems, such as grasses, cereals and canola (rape).

Heteroatom compounds: Chemical compounds that contain nitrogen and/or oxygen and/or sulfur and /or metals bound within their molecular structure(s).

Horizontal drilling: A drilling technique that permits the operator to contact and intersect a larger portion of the producing horizon than conventional vertical drilling techniques and can result in both increased production rates and greater ultimate recoveries of hydrocarbons.

Hydrocarbonaceous material: A material such as bitumen that is composed of carbon and hydrogen with other elements (heteroelements), such as nitrogen, oxygen, sulfur, and metals chemically combined within the structures of the constituents; even though carbon and hydrogen may be the predominant elements, there may be very few true hydrocarbons.

Hydrocarbon compounds: Chemical compounds containing only carbon and hydrogen.

Hydrodesulfurization: The removal of sulfur by hydrotreating (q.v.).

Hydroprocesses: Refinery processes designed to add hydrogen to various products of refining.

Hydrotreating: The removal of heteroatomic (nitrogen, oxygen, and sulfur) species by treatment of a feedstock or product at relatively low temperatures in the presence of hydrogen.

I

Idle cropland: Land in which no crops were planted; acreage diverted from crops to soil-conserving uses (if not eligible for and used as cropland pasture) under federal farm programs is included in this component.

Incinerator: Any device used to burn solid or liquid residues or wastes as a method of disposal.

Inclined grate: A type of furnace in which fuel enters at the top part of a grate in a continuous ribbon, passes over the upper drying section where moisture is removed, and descends into the lower burning section. Ash is removed at the lower part of the grate.

Incremental ultimate recovery: The difference between the quantity of oil that can be recovered by EOR methods and the quantity of oil that can be recovered by conventional recovery methods.

Indirect-injection engine: An older model of diesel engine in which fuel is injected into a pre-chamber, partly combusted, and then sent to the fuel-injection chamber.

Indirect liquefaction: Conversion of biomass to a liquid fuel through a synthesis gas intermediate step.

Industrial wood: All commercial round wood products except fuel wood.

Isopach: a line on a map designating points of equal formation thickness.

J

Jet fuel: Fuel meeting the required properties for use in jet engines and aircraft turbine engines.

Joule: Metric unit of energy, equivalent to the work done by a force of one Newton applied over distance of one meter (= 1 kg m^2/s^2). One joule (J) = 0.239 calories (1 calorie = 4.187 J).

K

Kerosene: A light middle distillate that in various forms is used as aviation turbine fuel or for burning in heating boilers or as a solvent, such as white spirit.

Kilowatt: (kW): A measure of electrical power equal to 1,000 watts. 1 kW = 3412 Btu/hr = 1.341 horsepower.

Kilowatt hour - (kWh): A measure of energy equivalent to the expenditure of one kilowatt for one hour. For example, 1 kWh will light a 100-watt light bulb for 10 hours. 1 kWh = 3412 Btu.

Kriging: A technique used in reservoir description for interpolation of reservoir parameters between wells based on random field theory.

L

Landfill gas: A type of biogas that is generated by decomposition of organic material at landfill disposal sites. Landfill gas is approximately 50 percent methane. See also biogas.

Leaded gasoline: Gasoline containing tetraethyl lead or other organometallic lead antiknock compounds.

Light petroleum: Petroleum having an API gravity greater than 20°.

Lignin: Structural constituent of wood and (to a lesser extent) other plant tissues, which encrusts the walls and cements the cells together.

Ligroine (Ligroin): A saturated petroleum naphtha boiling in the range of 20 to 135°C (68 to 275°F) and suitable for general use as a solvent; also called benzine or petroleum ether.

Liquefied petroleum gas: Propane, butane, or mixtures thereof, gaseous at atmospheric temperature and pressure, held in the liquid state by pressure to facilitate storage, transport, and handling.

Live cull: A classification that includes live cull trees; when associated with volume, it is the net volume in live cull trees that are 5.0 inches in diameter and larger.

Live steam: Steam coming directly from a boiler before being utilized for power or heat.

Logging residues: The unused portions of growing-stock and non-growing-stock trees cut or killed logging and left in the woods.

Lube: See Lubricating oil.

Lube cut: A fraction of crude oil of suitable boiling range and viscosity to yield lubricating oil when completely refined; also referred to as lube oil distillates or lube stock.

Lubricating oil: A fluid lubricant used to reduce friction between bearing surfaces.

M

M85: An alcohol fuel mixture containing 85 percent methanol and 15 percent gasoline by volume. Methanol is typically made from natural gas, but can also be derived from the fermentation of biomass.

Marsh: An area of spongy waterlogged ground with large numbers of surface water pools. Marshes usually result from: (1) an impermeable underlying bedrock; (2) surface deposits of glacial boulder clay; (3) a basin-like topography from which natural drainage is poor; (4) very heavy rainfall in conjunction with a correspondingly low evaporation rate; (5) low-lying land, particularly at estuarine sites at or below sea level.

Mbbls: 1,000 barrels of oil.

Mcf: 1,000 cubic feet of natural gas.

Mcfe: 1,000 cubic feet equivalent, determined using the ratio of six Mcf of natural gas to one barrel of crude oil.

Megawatt: (MW) A measure of electrical power equal to one million watts (1,000 kW).

MEOR: Microbial enhanced oil recovery.
Methanol: A fuel typically derived from natural gas, but which can be produced from the fermentation of sugars in biomass.
Microcrystalline wax: Wax extracted from certain petroleum residua and having a finer and less apparent crystalline structure than paraffin wax.
Middle distillate: Distillate boiling between the kerosene and lubricating oil fractions.
Migration (primary): The movement of hydrocarbons (oil and natural gas) from mature, organic-rich source rocks to a point where the oil and gas can collect as droplets or as a continuous phase of liquid hydrocarbon.
Migration (secondary): The movement of the hydrocarbons as a single, continuous fluid phase through water-saturated rocks, fractures, or faults followed by accumulation of the oil and gas in sediments (traps, q.v.) from which further migration is prevented.
Million: 1×10^6.
Mill residue: Wood and bark residues produced in processing logs into lumber, plywood, and paper.
Mineral hydrocarbons: Petroleum hydrocarbons, considered *mineral* because they come from the earth rather than from plants or animals.
Mineral oil: The older term for petroleum; the term was introduced in the 19th century as a means of differentiating petroleum (rock oil) from whale oil which, at the time, was the predominant illuminant for oil lamps.
Minerals: Naturally occurring inorganic solids with well-defined crystalline structures.
MMbbls: 1,000,000 barrels of oil.
MMBOE: 1,000,000 barrels of oil equivalent.
MMcf: 1,000,000 cubic feet of natural gas.
MMcfe: 1,000,000 cubic feet of gas equivalent, determined using the ratio of 6 Mcf of natural gas to 1 bbl of crude oil, condensate or natural gas liquids.
Modified/unmodified diesel engine: Traditional diesel engines must be modified to heat the oil before it reaches the fuel injectors in order to handle straight vegetable oil. Modified, any diesel engine can run on veggie oil; without modification, the oil must first be converted to biodiesel.

Moisture content: (MC): The weight of the water contained in wood, usually expressed as a percentage of weight, either oven-dry or as received.

Moisture content, dry basis: Moisture content expressed as a percentage of the weight of oven-wood, i.e.: [(weight of wet sample - weight of dry sample) / weight of dry sample] × 100.

Moisture content, wet basis: Moisture content expressed as a percentage of the weight of wood as-received, i.e.: [(weight of wet sample - weight of dry sample) / weight of wet sample] × 100.

MTBE: Methyl tertiary butyl ether is highly refined high octane light distillate used in the blending of petrol.

N

Naft: Pre-Christian era (Greek) term for naphtha (q.v.).

Napalm: A thickened gasoline used as an incendiary medium that adheres to the surface it strikes.

Naphtha: A generic term applied to refined, partly refined, or unrefined petroleum products and liquid products of natural gas, the majority of which distills below 240°C (464°F); the volatile fraction of petroleum which is used as a solvent or as a precursor to gasoline.

Native asphalt: See Bitumen.

Natural asphalt: See Bitumen.

Natural gas: The naturally occurring gaseous constituents that are found in many petroleum reservoirs; also there are also those reservoirs in which natural gas may be the sole occupant.

Natural gas liquids (NGL): The hydrocarbon liquids that condense during the processing of hydrocarbon gases that are produced from oil or gas reservoir; see also Natural gasoline.

Natural gasoline: A mixture of liquid hydrocarbons extracted from natural gas (q.v.) suitable for blending with refinery gasoline.

Natural gasoline plant: A plant for the extraction of fluid hydrocarbon, such as gasoline and liquefied petroleum gas, from natural gas.

Net: When used with respect to acres or wells, refers to gross acres of wells multiplied, in each case, by the percentage working interest owned by a company, individual, trust, or foundation.

Netback: Profit.

Net production: Production that is owned by a company, individual, trust, or foundation, less royalties and production due others.

Nitrogen fixation: The transformation of atmospheric nitrogen into nitrogen compounds that can be used by growing plants.

Nitrogen oxides (NOx): Products of combustion that contribute to the formation of smog and ozone.

Non-asphaltic road oil: Any of the nonhardening petroleum distillates or residual oils used as dust layers. They have sufficiently low viscosity to be applied without heating and, together with asphaltic road oils (q.v.), are sometimes referred to as dust palliatives.

Non-forest land: Land that has never supported forests and lands formerly forested where use of timber management is precluded by development for other uses; if intermingled in forest areas, unimproved roads and non-forest strips must be more than 120 feet wide, and clearings, etc., must be more than 1 acre in area to qualify as non-forest land.

Non-attainment area: Any area that does not meet the national primary or secondary ambient air quality standard established (by the Environmental Protection Agency) for designated pollutants, such as carbon monoxide and ozone.

Non-industrial private: An ownership class of private lands where the owner does not operate wood processing plants.

No. 1 Fuel oil: Very similar to kerosene (q.v.) and is used in burners where vaporization before burning is usually required and a clean flame is specified.

No. 2 Fuel oil: Also called domestic heating oil; has properties similar to diesel fuel and heavy jet fuel; used in burners where complete vaporization is not required before burning.

No. 4 Fuel oil: A light industrial heating oil and is used where preheating is not required for handling or burning; there are two grades of No. 4 fuel oil, differing in safety (flash point) and flow (viscosity) properties.

No. 5 Fuel oil: A heavy industrial fuel oil that requires preheating before burning.

No. 6 Fuel oil: A heavy fuel oil and is more commonly known as Bunker C oil when it is used to fuel ocean-going vessels; preheating is always required for burning this oil.

O

Oil: Crude oil or condensate.

Oil from tar sand: Synthetic crude oil (q.v.).

Oil mining: Application of a mining method to the recovery of bitumen.

OOIP (Oil originally in place or Original oil in place): The quantity of petroleum existing in a reservoir before oil recovery operations begin.

Open-loop biomass: Biomass that can be used to produce energy and bioproducts even though it was not grown specifically for this purpose; include agricultural livestock waste, residues from forest harvesting operations and crop harvesting.

Operator: The individual, company, trust, or foundation responsible for the exploration, development, and production of an oil or gas well or lease.

Oxygenate: A substance which, when added to gasoline, increases the amount of oxygen in that gasoline blend; includes fuel ethanol, methanol, and methyl tertiary butyl ether (MTBE).

P

Particulate emissions: particles of a solid or liquid suspended in a gas, or the fine particles of carbonaceous soot and other organic molecules discharged into the air during combustion.

Particulate: A small, discrete mass of solid or liquid matter that remains individually dispersed in gas or liquid emissions.

Pay zone thickness: The depth of a tar sand deposit from which bitumen (or a product) can be recovered.

Permeability: The ease of flow of the water through the rock.

Petrol: A term commonly used in some countries for gasoline.

Petrolatum: A semisolid product, ranging from white to yellow in color, produced during refining of residual stocks; see Petroleum jelly.

Photosynthesis: Process by which chlorophyll-containing cells in green plants concert incident light to chemical energy, capturing carbon dioxide in the form of carbohydrates.

Pour point: The lowest temperature at which oil will pour or flow when it is chilled without disturbance under definite conditions.

Present value of future revenues: The pretax present value of estimated future revenues to be generated from the production of proved reserves, net of estimated production and future development costs. Future net revenues are discounted to a present value of an annual discount rate whcih is typically 10%.

Primary oil recovery: Oil recovery utilising only naturally occurring forces; recovery of crude oil from the reservoir using the inherent reservoir energy.

Primary wood-using mill: A mill that converts round wood products into other wood products; common examples are sawmills that convert saw logs into lumber and pulp mills that convert pulpwood round wood into wood pulp.

Process heat: Heat used in an industrial process rather than for space heating or other housekeeping purposes.

Producer gas: Fuel gas high in carbon monoxide (CO) and hydrogen (H2), produced by burning a solid fuel with insufficient air or by passing a mixture of air and steam through a burning bed of solid fuel.

Proved developed reserves: Reserves that can be expected to be recovered through existing wells with existing equipment and operating methods. Additional oil and gas expected to be obtained through the application of fluid injection or other improved recovery techniques for supplementing the natural forces and mechanisms of primary recovery can be included as "proved developed reserves" only after testing by a pilot project, or after the operation of an installed program has confirmed through production response that increased recovery will be achieved.

Proved reserves: The estimated quantities of crude oil, natural gas, and natural gas liquids which geological and engineering data demonstrate with reasonable certainty to be recoverable in future years from known reservoirs under existing economic and operating conditions, i.e., prices and costs as of the date the estimate is made; prices include consideration of changes in existing prices provided only by contractual arrangements, but not on escalations based upon future conditions.

Proved undeveloped reserves: Reserves that are expected to be recovered from new wells on undrilled acreage, or from existing wells where a relatively major expenditure is required for recompletion; reserves on undrilled acreage are usually limited to those drilling units offsetting productive units that are reasonably certain of production when drilled; proved reserves for other undrilled units can be claimed only where it can be demonstrated with certainty that there is continuity of production from the existing productive formation; under no circumstances are estimates for proved undeveloped reserves generally attributed to any acreage for which an application of fluid injection or other improved recovery technique is contemplated, unless such techniques have been proved effective by actual tests in the area and in the same reservoir.

Pulpwood: Round wood, whole-tree chips, or wood residues that are used for the production of wood pulp.

Pyrolysis: The thermal deçomposition of biomass at high temperatures (greater than 400°F, or 200°C) in the absence of air; the end product of pyrolysis is a mixture of solids (char), liquids (oxygenated oils), and gases (methane, carbon monoxide, and carbon dioxide) with proportions determined by operating temperature, pressure, oxygen content, and other conditions.

Q

Quad: One quadrillion Btu (10^{15} Btu) = 1.055 exajoules (EJ), or approximately 172 million barrels of oil equivalent.

Quadrillion: 1×10^{15}.

Quench: the sudden cooling of hot material discharging from a thermal reactor.

R

Raw materials: Minerals extracted from the earth prior to any refining or treating.

Recompletion: The completion for production of an existing well bore in another formation from that in which the well has been previously completed.

Recovery boiler: A pulp mill boiler in which lignin and spent cooking liquor (black liquor) is burned to generate steam.

Refining: The processes by which petroleum is distilled and/or converted by application of a physical and chemical processes to form a variety of products are generated.

Reformulated gasoline (RFG): Gasoline designed to mitigate smog production and to improve air quality by limiting the emission levels of certain chemical compounds such as benzene and other aromatic derivatives; often contains oxygenates (q.v.).

Refractory lining: A lining, usually of ceramic, capable of resisting and maintaining high temperatures.

Refuse-derived fuel (RDF): Fuel prepared from municipal solid waste; non-combustible materials such as rocks, glass, and metals are removed, and the remaining combustible portion of the solid waste is chopped or shredded.

Reserves: Proved reserves; well-identified resources that can be profitably extracted and utilized with existing technology.

Residues: Bark and woody materials that are generated in primary wood-using mills when round wood products are converted to other products.

Residual oil: Oil that cannot be recovered from a conventional oil reservoir after application of primary, secondary and enhanced oil recovery techniques.

Residuum (pl. residua, also known as resid or resids): The non-volatile portion of petroleum that remains as residue after refinery distillation; hence, atmospheric residuum, vacuum residuum.

Resource: The total amount of a commodity (usually a mineral but can include non-minerals such as water and petroleum) that has been estimated to be ultimately available.

Rock matrix: The granular structure of a rock or porous medium.

Rotation: Period of years between establishment of a stand of timber and the time when it is considered ready for final harvest and regeneration.

Round wood products: Logs and other round timber generated from harvesting trees for industrial or consumer use.

Royalty: An interest in an oil and gas lease that gives the owner of the interest the right to receive a portion of the production from the leased acreage (or of the proceeds of the sale of production), but generally does not require the owner to pay any portion of the costs of drilling or operating the wells on the leased acreage; royalties may be either landowner's royalties, which are reserved by the owner of the leased acreage at the time the lease is granted, or overriding royalties, which are usually reserved by an owner of the leasehold in connection with a transfer to a subsequent owner.

Run-of-the-river reservoirs: Reservoirs with a large rate of flow-through compared to their volume.

S

Sand: A course granular mineral mainly comprising quartz grains that is derived from the chemical and physical weathering of rocks rich in quartz, notably sandstone and granite.

Sandstone: A sedimentary rock formed by compaction and cementation of sand grains; can be classified according to the mineral composition of the sand and cement.

Saturated steam: Steam at boiling temperature for a given pressure.

Secondary oil recovery: Application of energy (e.g., water flooding) to recovery of crude oil from a reservoir after the yield of crude oil from primary recovery diminishes.

Secondary wood processing mills: A mill that uses primary wood products in the manufacture of finished wood products, such as cabinets, moldings, and furniture.

Sedimentary: Formed by or from deposits of sediments, especially from sand grains or silts transported from their source and deposited in water, as sandstone and shale; or from calcareous remains of organisms, as limestone.

Sedimentary strata: Typically consist of mixtures of clay, silt, sand, organic matter, and various minerals; formed by or from deposits of sediments, especially from sand grains or silts transported from their source and deposited in water, such as sandstone and shale; or from calcareous remains of organisms, such as limestone.

Shut-in royalty: A payment stipulated in the oil lease, which royalty owners receive in lieu of actual production, when a well is shut-in due to lack of a suitable market, a lack of facilities to produce the product, or other cases defined within the shut-in provisions contained in the oil and gas lease.

Slime: A name used for petroleum in ancient texts.

Sonic log: A type of acoustic log that displays travel time of P-waves versus depth; sonic logs are typically recorded by pulling a tool on a wireline up the wellbore; the tool emits a sound wave that travels from the source to the formation and back to a receiver.

Specific gravity: The mass (or weight) of a unit volume of any substance at a specified temperature compared to the mass of an equal volume of pure water at a standard temperature.

Stand (of trees): A tree community that possesses sufficient uniformity in composition, constitution, age, spatial arrangement, or condition to be distinguishable from adjacent communities.

Steam turbine: A device for converting energy of high-pressure steam (produced in a boiler) into mechanical power, which can then be used to generate electricity.

Straight vegetable oil (SVO): Any vegetable oil that has not been optimized through the process of transesterification.

Strata: Layers including the solid iron-rich inner core, molten outer core, mantle, and crust of the earth.

Superheated steam: Steam that is hotter than boiling temperature for a given pressure.

Surfactant: A type of chemical, characterized as one that reduces interfacial resistance to mixing between oil and water or changes the degree to which water wets reservoir rock.

Sustainable: An ecosystem condition in which biodiversity, renewability, and resource productivity are maintained over time.

Synthetic crude oil (syncrude): A hydrocarbon product produced by the conversion of coal, oil shale, or tar sand bitumen that resembles conventional crude oil; can be refined in a petroleum refinery.

Synthetic ethanol: Ethanol produced from ethylene, a petroleum by-product.

T

Tar sand (bituminous sand): A formation in which the bituminous material (bitumen) is found as a filling in veins and fissures in fractured rocks or impregnating relatively shallow sand, sandstone, and limestone strata; a sandstone reservoir that is impregnated with a heavy, extremely viscous, black hydrocarbonaceous, petroleum-like material that cannot be retrieved through a well by conventional or enhanced oil recovery techniques; (FE 76-4): The several rock types that contain an extremely viscous hydrocarbon which is not recoverable in its natural state by conventional oil well production methods including currently used enhanced recovery techniques.

Tertiary recovery: Enhanced recovery methods for the production of oil or gas; enhanced recovery of crude oil requires a means for displacing oil from the reservoir rock, modifying the properties of the fluids in the reservoir and/or the reservoir rock to cause movement of oil in an efficient manner, and providing the energy and drive mechanism to force its flow to a production well; chemicals or energy is injected as required for displacement and for the control of flow rate and flow pattern in the reservoir, and a fluid drive is provided to force the oil toward a production well.

Thermochemical conversion: Use of heat to chemically change substances from one state to another, e.g. to make useful energy products.

Timberland: Forest land that is producing or is capable of producing crops of industrial wood, and that is not withdrawn from timber utilization by statute or administrative regulation.

Tipping fee: A fee for disposal of waste.

Ton (short ton): 2,000 pounds.

Tonne (Imperial ton, long ton, shipping ton): 2,240 pounds; equivalent to 1,000 kilograms or in crude oil terms about 7.5 barrels of oil.

Topping cycle: A cogeneration system in which electric power is produced first. The reject heat from power production is then used to produce useful process heat.

Topping and back pressure turbines: Turbines which operate at exhaust pressure considerably higher than atmospheric (non-condensing turbines); often multistage with relatively high efficiency.

Transesterification: The chemical process in which an alcohol reacts with the triglycerides in vegetable oil or animal fats, separating the glycerin and producing biodiesel.

Traveling grate: A type of furnace in which assembled links of grates are joined together in a perpetual belt arrangement. Fuel is fed in at one end and ash is discharged at the other.

Trillion: 1×10^{12}.

Turbine: A machine for converting the heat energy in steam or high temperature gas into mechanical energy. In a turbine, a high velocity flow of steam or gas passes through successive rows of radial blades fastened to a central shaft.

Turn down ratio: The lowest load at which a boiler will operate efficiently as compared to the boiler's maximum design load.

U

Ultimate recovery: The cumulative quantity of oil that will be recovered when revenues from further production no longer justify the costs of the additional production.

Unconformity: A surface of erosion that separates younger strata from older rocks.

Upgrading: The conversion of petroleum to value-added salable products.

V

Vacuum distillation: A secondary distillation process which uses a partial vacuum to lower the boiling point of residues from primary distillation and extract further blending components.

Viscosity: A measure of the ability of a liquid to flow or a measure of its resistance to flow; the force required to move a plane surface of area 1 square meter over another parallel plane surface 1 meter away at a rate of 1 meter per second when both surfaces are immersed in the fluid; the higher the viscosity, the slower the liquid flows.

Volatile Organic Compounds (VOCs): Name given to light organic hydrocarbons which escape as vapor from fuel tanks or other sources, and during the filling of tanks. VOCs contribute to smog.

Volumetric sweep: The fraction of the total reservoir volume within a flood pattern that is effectively contacted by injected fluids.

VSP: Vertical seismic profiling, a method of conducting seismic surveys in the borehole for detailed subsurface information.

W

Waste streams: Unused solid or liquid by-products of a process.

Waste vegetable oil (WVO): Grease from the nearest fryer which is filtered and used in modified diesel engines, or converted to biodiesel through the process of transesterification and used in any ol' diesel car.

Water-cooled vibrating grate: A boiler grate made up of a tuyere grate surface mounted on a grid of water tubes interconnected with the boiler circulation system for positive cooling; the structure is supported by flexing plates allowing the grid and grate to move in a vibrating action; ash is automatically discharged.

Watershed: The drainage basin contributing water, organic matter, dissolved nutrients, and sediments to a stream or lake.

Watt: The common base unit of power in the metric system; one watt equals one joule per second, or the power developed in a circuit by a current of one ampere flowing through a potential difference of one volt. One Watt = 3.412 Btu/hr.

Wellbore: The hole in the earth comprising a well.

Well completion: The complete outfitting of an oil well for either oil production or fluid injection; also the technique used to control fluid communication with the reservoir.

Wellhead: That portion of an oil well above the surface of the ground.

Wettability: The relative degree to which a fluid will spread on (or coat) a solid surface in the presence of other immiscible fluids.

Wettability number: A measure of the degree to which a reservoir rock is water-wet or oil-wet, based on capillary pressure curves.

Wettability reversal: The reversal of the preferred fluid wettability of a rock, e.g., from water-wet to oil-wet, or vice versa.

Wheeling: The process of transferring electrical energy between buyer and seller by way of an intermediate utility or utilities.

Whole-tree harvesting: A harvesting method in which the whole tree (above the stump) is removed.

Working interest: An interest in an oil and gas lease that gives the owner of the interest the right to drill for and produce oil and gas on the leased acreage and requires the owner to pay a share of the costs of drilling and production operations; the share of production to which a working interest owner is entitled will always be smaller than the share of costs that the working interest owner is required to bear, with the balance of the production accruing to the owners of royalties; thus, the owner of a 100% working interest in a lease burdened by a landowner's royalty of 12.5% would be required to pay 100% of the costs of a well but would be entitled to retain 87.5% of the production.

Workover: Operations on a producing well to restore or increase production.

Y

Yarding: The initial movement of logs from the point of felling to a central loading area or landing.

Z

Zeolite: a crystalline aluminosilicate used as a catalyst and having a particular chemical and physical structure.

Index

Also of Interest

Check out these other related titles from Scrivener Publishing

Now Available from the Same Author:
Ethics in Engineering, by James Speight and Russell Foote, ISBN 9780470626023. Covers the most thought-provoking ethical questions in engineering.

Advances in Natural Gas Engineering Series (3 volumes):
Acid Gas Injection and Related Technologies, by Ying Wu and John J. Carroll, ISBN 9781118016640. The only book covering this timely topic in natural gas. *NOW AVAILABLE!*

Carbon Dioxide Sequestration and Related Technologies, by Ying Wu and John J. Carroll, ISBN 9780470938768. Volume two focuses on one of the hottest topics in any field of engineering, carbon dioxide sequestration. *NOW AVAILABLE!*

Sour Gas and Related Technologies, by Ying Wu and John J. Carroll, ISBN 9780470948149. This third volume in the series focuses on sour gas, one of the most important issues facing chemical and process engineers. *PUBLISHING MAY 2012*

Other books in Energy and Chemical Engineering:
Emergency Response Management for Offshore Oil Spills, by Nicholas P. Cheremisinoff, PhD, and Anton Davletshin, ISBN 9780470927120. The first book to examine the Deepwater Horizon disaster and offer processes for safety and environmental protection. *NOW AVAILABLE!*

Advanced Petroleum Reservoir Simulation, by M.R. Islam, S.H. Mousavizadegan, Shabbir Mustafiz, and Jamal H. Abou-Kassem, ISBN 9780470625811. The state of the art in petroleum reservoir simulation. *NOW AVAILABLE!*

Energy Storage: A New Approach, by Ralph Zito, ISBN 9780470625910. Exploring the potential of reversible concentrations cells, the author of this groundbreaking volume reveals new technologies to solve the global crisis of energy storage. *NOW AVAILABLE!*

Zero-Waste Engineering, by Rafiqul Islam, February 2012, ISBN 9780470626047. In this controvercial new volume, the author explores the question of zero-waste engineering and how it can be done, efficiently and profitably.

Fundamentals of LNG Plant Design, by Saeid Mokhatab, David Messersmith, Walter Sonne, and Kamal Shah, August 2012. The only book of its kind, detailing LNG plant design, as the world turns more and more to LNG for its energy needs.

Flow Assurance, by Boyun Guo and Rafiqul Islam, January 2012, ISBN 9780470626085. Comprehensive and state-of-the-art guide to flow assurance in the petroleum industry.

Printed in the United States
By Bookmasters